W9-DIF-211

TELECOM BASICS

by JACK L. DEMPSEY

Telephony

Div. Intertec Publishing Corp.
55 E. Jackson Blvd.
Chicago, IL 60604

First Edition: February 1988

Library of Congress Catalog Card Number: 87-83662

ISBN: 0-917845-07-2

Printed in the United States of America

PREFACE

A basic knowledge of telecommunications is essential to anyone entering the field—from a newly hired employee to an executive assigned to a position related to telecommunications. In the past, it was not unusual for individuals to gradually obtain such knowledge through osmosis. Today, the pace is so rapid and so many firms are in telecommunications-related fields, that there is a demand for virtually instant knowledge.

This book provides fundamentals on almost all aspects of telecommunications. The reader will be well-equipped to enter into or expand his grasp of this complex field. He will attain a comfortable speaking knowledge of all elementary facets of voice and data communications, ranging from such basics as to how a telephone instrument works to such esoteric topics as electronic tandem networks or statistical multiplexers. All of them are presented in a down-to-earth, easy-to-grasp manner with profuse illustration.

Jack L. Dempsey
St. Petersburg, Florida
Jan., 1988

TABLE OF CONTENTS

1 FUNDAMENTALS

A Brief History As everyone knows, the telephone was invented by Alexander Graham Bell ... but was it really? Yes, but another man, Elisha Gray, *simultaneously* developed a similar device. In fact he applied for his patent just a few hours after Bell did.

Initially, Bell tried to sell his patent to the then giant Western Union Company. The leaders of the company saw no practical use for the device. Sound familiar? Have you heard the story of Chester Carlson and the Xerox machine? Carlson went to the giants such as IBM and General Electric trying in vain to sell his concept of a dry copying device. Their attitude was that carbon paper worked just fine.

Bell managed to convince enough backers that his invention had a future, and he formed the American Telephone and Telegraph Company. Later on, AT&T could have tucked Western Union into its pocket.

HOW A PHONE WORKS, A PRIMER

The first commercial phone consisted of a single unit that was used both to speak into and to listen with. Convenience dictated the addition of a second identical unit so that the user could talk and listen at the same time. Even today, it is possible to talk into the listening end, the receiver, of a phone. Perhaps the word, "shout", should be used in place of "talk" ... and that was the problem, the receiver was inefficient as a microphone.

Receiver Let us then examine the receiver to see how it works. Refer to Figure 1.1. The same concept was used in the very earliest receivers. At the heart of a telephone receiver (speaker) there is a magnet. Notice that there is a coil of wire shown wound around the permanent magnet, so that the magnet is both a permanent one and an electromagnet at the same time.

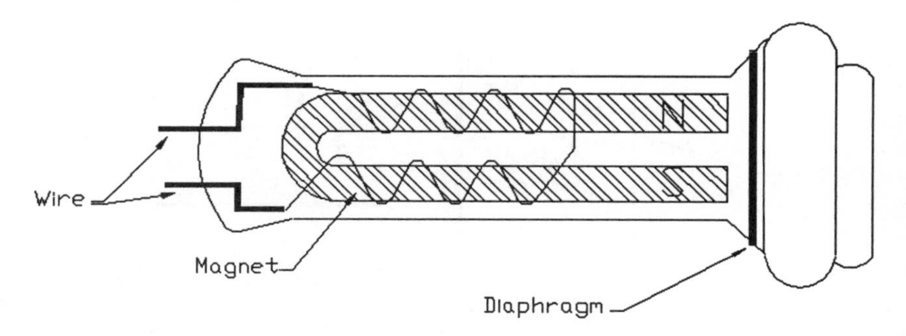

Figure 1.1 Receiver

Notice also the diaphragm that is close to the face of the magnet. It is designed so it doesn't quite touch the face. Imagine now that a current that is varying in response to voice frequencies is applied to the coil of wire. It can be envisioned that the magnetism will vary with the amount of instantaneous current, and that this, in turn, will cause the diaphragm to vibrate in concert with the varying current. This is the essence of a telephone receiver.

Interestingly enough, if a person were to shout into the receiver, it would generate a current in the coil wrapped around the magnet that is in concert with the speech energy. So, we see that the receiver can double as a microphone, or transmitter, as it is called in telephony.

Transmitter Thomas Edison got involved in the telephone business, too. He invented the carbon transmitter. This transmitter (microphone) is the type used in the vast majority of the telephones today.

The carbon transmitter takes advantage of the fact that carbon, the material "lead" pencils are made of, has an interesting property. When carbon is put under pressure, its electrical resistance decreases in proportion to the amount of pressure. Of course, the opposite is also true. If the resistance decreases, then more electrical current can flow.

Let us now look at Figure 1.2. This figure is a simplified diagram of a telephone transmitter. A flexible metal diaphragm is in close contact with carbon particles. An electrical circuit is arranged so that a current is flowing through the carbon. Now, if someone should speak into the transmitter, the sound waves will cause the diaphragm to vibrate in concert with the voice. The vibrating diaphragm alternately compresses and decompresses the carbon. The result is that the electrical current increases and decreases in the same way. Hence, the transmitter has converted voice energy into electrical energy.

Telephone Bell As useful as a telephone is, it is virtually worthless without some way to let a called party know that there is a call for him. Initially, this was a problem. The trick was to use the same wires to talk

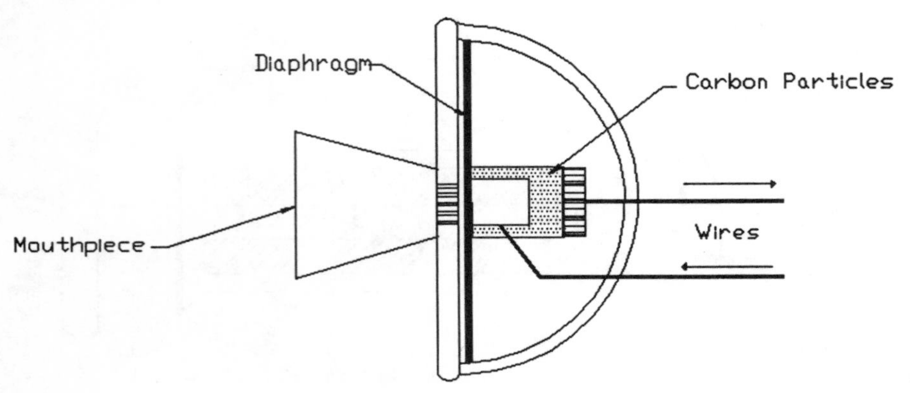

Figure 1.2 Transmitter

and signal. It was solved by the use of a hand-cranked magneto. A magneto puts out an alternating current that can be separated from the direct current that a phone uses for communications.*

Rotary Dial The author is old enough to remember the days when most phones did not have a dial, even though the concept of automatic switching by dial was invented just ten years after the telephone itself. So-called *manual* service was the most common until the 1950s. When the phone was lifted off the hook, an operator would say, "number please?" However, waiting for her to serve a subscriber could take up to a minute or more. She established all connections at a cord-equipped switchboard.

An undertaker named Strowger invented the first dial system. It seems that he thought that the local operator was sending calls that were intended for him to his competitor across town. Determined to develop a system that would eliminate such acts, he invented the so-called step-by-step dial system. (It is still often called the "Strowger" system.)

Looking at Figure 1.3, we see that the dial produces interruptions, called "pulses", in the current of the telephone circuit . Dialing a "5" produces five pulses, etc. The switching equipment responds to the pulses. More on this later.

Tone Dial Tone dialing was no doubt inspired by the arrival of the "pushbutton" age after World War II. As early as the 1950s, Bell Labs had a trial in Elgin, Illinois, of a rather crude pushbutton dialer as it was called. This early dialing device used a system of reeds that were "twanged" when a button was pushed. The concept was adequate, but unreliable. The idea had to wait for inexpensive and reliable transistors in the 60s for it to become viable.

However, tone dialing was already in place and working well on long distance (toll) circuits. The *multifrequency* (MF) signaling system was and is still used by operators and switches for intercity circuit call setup. It laid the basis for subscriber tone dialing. The most significant reason why

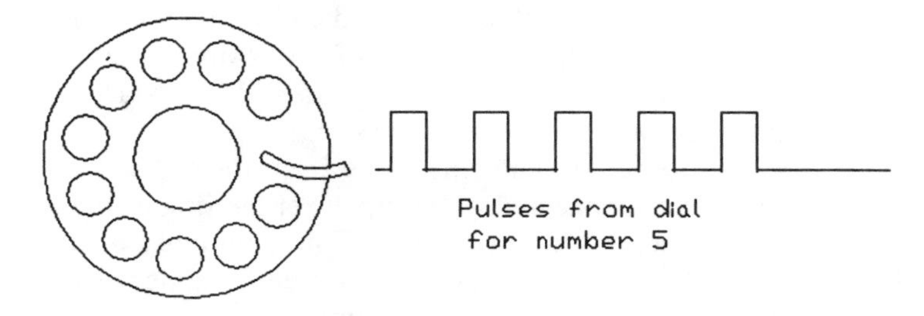

Figure 1.3
Dial Pulsing

*Telephone bells generally are rung with 20-cycle current rather than the 60-cycle that is common in household lighting. The reason is that the crank could only be conveniently turned fast enough to produce about 20 cycles per second.

MF could be used for toll call setup was because vacuum tubes could be used to produce the tones. Tubes are out of the question for normal telephone sets.

Now that user tone dialing is commonplace, almost everyone recognizes that when a button is pressed two tones (not in harmony) are produced. Each digit and special characters * and # has its own set of two tones. (Incidentally, for readers interested in trivia, the "#" is properly called an "octathorp".) The Bell System coined the registered name of "Touch Tone" for the service. It is more properly called DTMF, an abbreviation for "Dual Tone Multi-frequency". The tones selected were not arrived at capriciously. They are all in the center of the normal telephone circuit frequency band. Also, any distortion present in the circuits will not artificially cause some tones to falsely produce another legitimate one.

Frequency	1209	1336	1477	1633
697	1	2	3	A
770	4	5	6	B
852	7	8	9	C
941	*	0	#	D

Tone Dialing Arrangement

There is actually a total of eight tones, arranged as shown in the matrix. Pressing any of the normal 12 buttons causes the two tones depicted at the intersection to sound. Note that there is a possibility for 16 buttons. The extra four are used in the military *Autovon* system.

Tip and Ring It is likely that everyone associated with the telephone business in North America is familiar with the term, "tip and ring". This is the name used to designate the two wires of a normal telephone line. First of all, why is it necessary to distinguish between the two? In days past, it was necessary to keep the wires differentiated because most people had party lines, and ringing current was applied to one or the other of the wires to provide selective ringing. Because of the many electronic devices such as speakerphones and tone dialers, it is still necessary to keep the two wires differentiated for them to work.

Refer to Figure 1.4. The terms, tip and ring come from the old switchboards. The tip of the plug was connected to the tip wire; the ring of brass just behind the tip was connected to the ring wire. The main body of the plug is called the "sleeve". The sleeve conductor was used for supervisory purposes; that is, as a control over the switching connection.

Network A group of components is used to make the transmitter and receiver match the characteristics of line that the phone is connected to. Taken together, these parts are called the "network." A phone will work

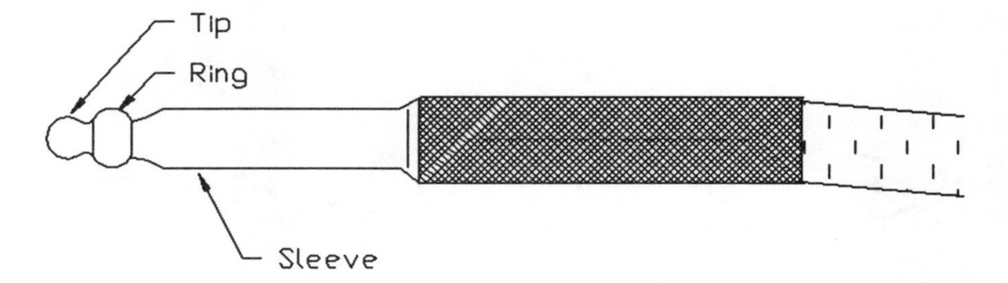

Figure 1.4
Switchboard Plug

without a network, but not as well; so all commercial phones have them.

This, then, is a basic telephone—a rather remarkable instrument that has changed only a little through the years. Except for cheap imports, it is amazingly rugged and dependable. Electronics have been creeping into its design in the form of tone dials, tone ringers and improved transmitters at a slow pace. Perhaps this is a continuing testimonial to the excellence of the original design.

THE NEED FOR SWITCHING

Obviously, telephones would be worthless from a practical standpoint without an efficient way to connect them together on demand. This was recognized from the beginning, however, in the earliest days establishing the desired connection was left to the owner of an instrument. In fact, running the wires was left to the owner, too. Crude switches on the premises were used to connect to the proper set of wires. As can be seen in Figure 1.5, a maze of wires soon results when as few as six locations are interconnected. Imagine what would have to be done if someone else wanted to join the network? Mathematically, the following expression shows that the number of interconnections increases astronomically:

$$I = N(N-1)/2$$

Where: I is the number of interconnections
N is the number of subscribers to the service

If there were just 50 subscribers, the number of interconnections would have to be 50 x 49/2, or 1225 sets of wires going to each location.

Clearly, this arrangement is not workable. This brings us to the need for switching at a central location, or a "central office".

CENTRAL OFFICE

A central office (CO) is nothing more than a hubbing place where all users' phone lines are brought together and a means of interconnecting any one to any other one is provided. A very important point, though, is that means are *not* provided to simultaneously attend to all users at once. In other words, there is a definite *probability of blocking* on a given

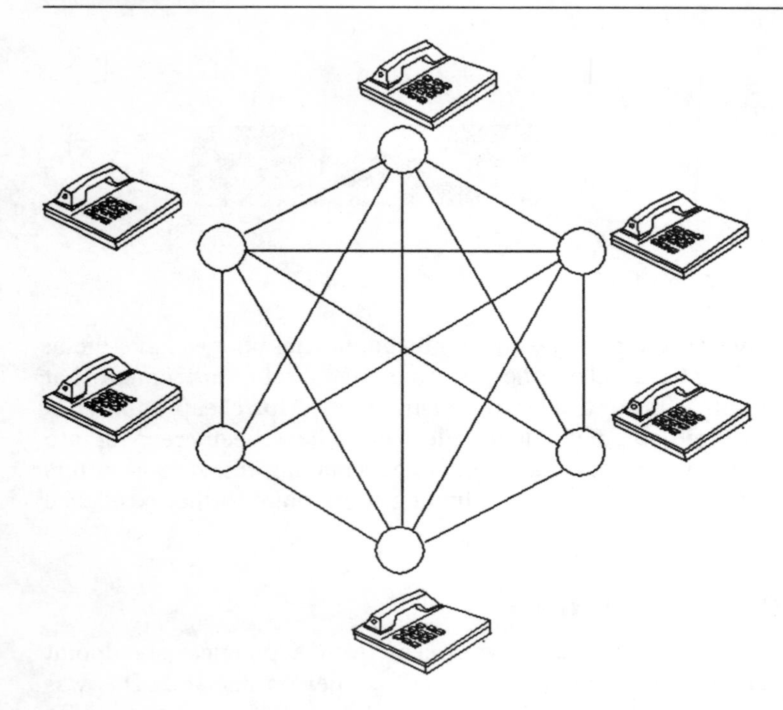

Figure 1.5
Six Points Interconnected

attempt to complete a call. More on this when the topic of *traffic* is covered later.

At first, boys were used to manually handle calls. It was soon discovered that they were prone to use bad language with the customers. This provided one of the first entres into the business world for women, who were more likely to be soft-spoken and gentle.*

COs range from manual to the most complex stored-program electronic offices. Because almost all types of COs have counterparts in Private Branch Exchanges (PBXs) which will be discussed later, we will not go into detail on the operation of COs.

Early in the history of telephony, only local calling was possible. A central office was like an island, serving only the local users. As long distance service evolved, means of interconnecting COs were provided using *trunks* between the offices. Later we will examine the hierarchy of trunks and the various types of offices that were established to handle long distance service.

*If dial systems were not in place today, there would never be enough people in the work force to be operators, male or female, to begin to handle all of today's calling.

2 TRANSMISSION I

FREQUENCY

Frequency refers to the number of times per second that an object vibrates, or voltage changes or current changes. In the days when meaningful terms were still used, frequency was expressed in *cycles per second*. Today it is only proper to refer to values of frequency as being in hertz. A cycle-per-second and a hertz are the same.

The lowest frequency encountered in telephony is on or about 20 hertz, the frequency of ringing current. However, frequencies go much much higher in telephony: microwave transmission systems use frequencies in *billions* of hertz.

Voice Frequencies Of particular interest to us humans is the range of frequencies that we can hear: 20 to 20,000 hertz. This is the range of a good stereo system. It happens, however, that intelligible and recognizable speech can be transmitted in a 300 to 3300 hertz range. This is indeed fortunate since this is just about the limit of the inexpensive transmitters and receivers used in phones. A *bandwidth* of 3000 hertz (3300-300) is all that is needed to carry speech.

Voice signals are called analog as opposed to digital. That is, they are smoothly varying such as a sine wave shown in Figure 2.1.

Amplitude Amplitude is another way of saying how big a thing is. Usually in telephony we are referring to the *instantaneous* amount of voltage or power a signal has. Referring to Figure 2.1, we see that the amplitude of a sine wave varies smoothly with time. If we were to listen to such a sine wave signal it would sound like a constant tone of a certain

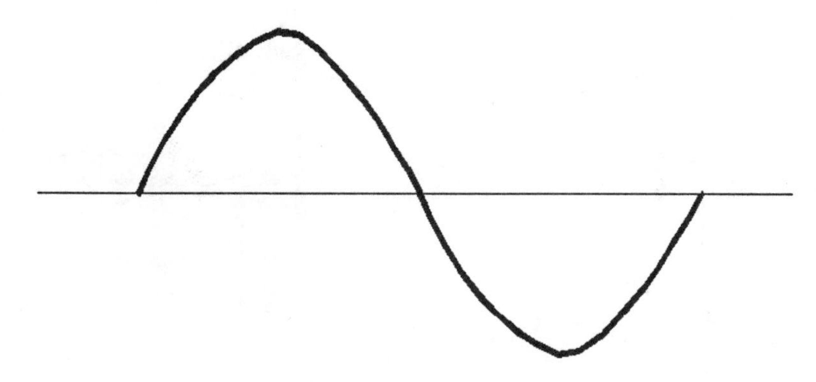

Figure 2.1 Sine Wave

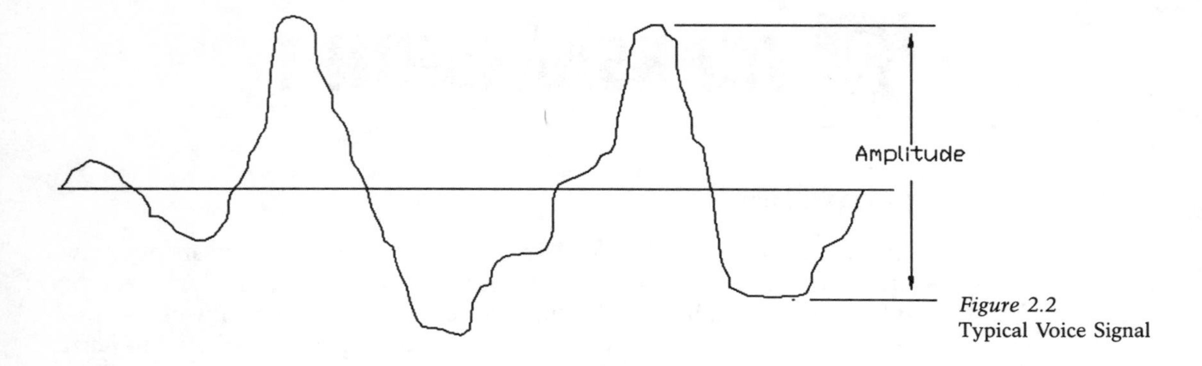

Figure 2.2
Typical Voice Signal

frequency. Of greater interest is Figure 2.2. Here we see a typical voice signal. Notice that the amplitude variations are not smooth, although there is some predictability about them; there are no wild swings such as would be heard with noise or rock music.

Later, we will see how amplitude variations can be used to carry more than one voice signal at a time over wires or other media.

TWO-WIRE CIRCUITS

This section is headed, "Two-wire Circuits" because it is possible to carry speech signals on one wire or four, depending on how it is done. Since telegraph was an established service at the time of the invention of the telephone, it is not surprising that the techniques used in telegraphy were carried over to telephony. Also, burglar alarm companies provided many of the circuits, and they used single wires and ground return. Thus, early telephone transmission was handled by "one-wire circuits". High school science classes taught us that a circuit requires at least two wires. How does a one-wire "circuit" work? Easy. Earth or ground is used to complete the circuit. See Figure 2.3.

Streetcars were the demise of such circuits for telephony. The electric

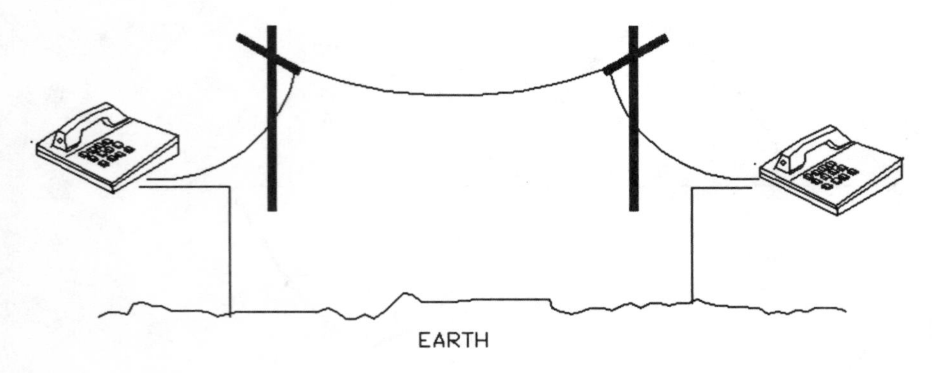

Figure 2.3
A "One-Wire" Connection

circuits used to provide power to them depended on the use of ground as a circuit return. Of course the same ground was shared by the phone wires; streetcar currents could be heard in the form of loud noise on the phone circuit. With telegraphy, the noise wasn't great enough to affect the ON and OFF (binary) telegraph signals. This was really too bad. A lot of wire could have been saved.

The first telephone conversation is reported not to have been a "conversation" at all, but a one-way command by Dr. Bell to his assistant to come to his aid. He is said to have shouted, "Mr. Watson. Come here, I want you!" Barring this case, the early telephone calls were two-way affairs as they are today; however, the early calls were always over a pair of wires, once it was realized that the use of single wires was impractical.

FOUR-WIRE CIRCUITS

Since telephone calls can obviously be carried on two wires, why are four wires so frequently used when it appears that it doubles the cost?

Despite the perception that using four wires costs a lot more than using two, it doesn't work out that way when the circuit is more than a few miles long. And, as is so often the case, the payback of the additional cost is great. A higher quality connection is the result. The reason for this is that amplifiers are needed to boost the signal to make up for losses that occur as function of distance, and a four-wire circuit permits the use of much simpler, greater powered and more stable amplifiers. More on this in the section on repeaters.

There is another more compelling reason that all but the the very shortest circuits are four wire. Carrier systems (to be discussed later) are used to provide longer-haul circuits, and they are inherently four-wire devices—that is, there is a circuit path for each direction of transmission.

Hybrids/Four-Wire Termination The problem in using four-wire circuits is where "four become two". When circuits are merged, there is a possibility of howling if there isn't a good match, (much like a marriage).

In order to overcome circuit losses, it is necessary to insert amplification in each of the two legs of a four-wire circuit. Refer to Figure 2.4. Unless something is done to prevent it, the energy put into the East end of the circuit can be fed back at the West end junction right back to the East end. It then can go whipping back and forth. The result, at a minimum, is a "rainbarrel" effect; at worse, a circuit howl.

The clue to making the two- to four-wire conversion work is "matching" in devices called *hybrids,* or sometimes, *four-wire term sets.* (Electrical engineers will recognize the hybrid as a special form of a bridge circuit.) Stated simply, a hybrid that is well matched to the line is like a traffic cop who keeps cars moving in the right direction. A hybrid and associated balancing networks are designed to keep most of the incoming signal energy from returning to the other end of the circuit. Matching the

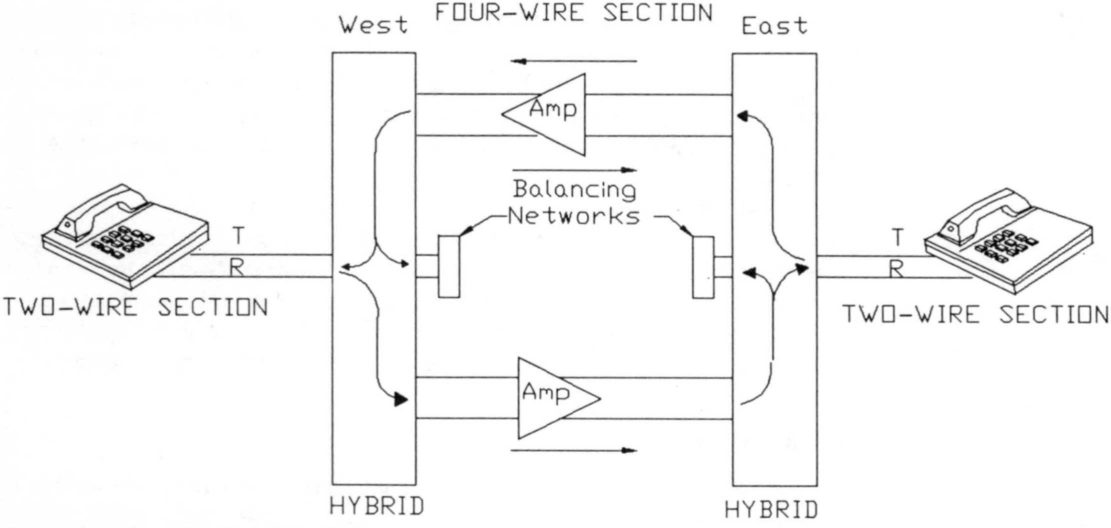

Figure 2.4 Two-Wire/Four-Wire System

network to the line is a problem at times, and requires detailed engineering and consideration of the circuits' parameters.

The introduction of carrier systems (microwave, for example,) to telephone plant resulted in virtually all long haul circuits being provided by means of four-wire equivalent circuits, since they are intrinsically devices that use two separate paths.

Repeaters The word, "repeater" is telephonese for amplifier. The term originated with telegraphy, where a device "repeated" the weak on-and-off pulses of electrical current when they became weak along the lines. Electro-mechanical "relays" followed the weak incoming current and produced brand new, strong pulses to send along the line.

Relays worked well with telegraphy, where only pulses had to be reproduced. But what of voice signals, which aren't pulses? (That is, unless they are "digitized"—to be covered later in this book.) As was noted above, voice signals are analog in nature.*

The solution to the problem caused a lot of head-scratching among the telephone engineers. The early circuits were only two-wire facilities, and the means to provide amplification in both directions at once was and is a very difficult engineering problem.

The way to a solution was paved by the use of four-wire circuits, as

*At first, the problem of boosting weak voice signals along the way was solved by means of human "repeaters". Literally, people were stationed along the way on a long distance call to listen and repeat what they heard! Obviously, this technique was doomed to early replacement.

noted above. If the circuits could be separated by direction of transmission, all that was needed was to amplify the signals in each direction.*

Repeaters generally are placed along a telephone line every 40 or 50 miles to make up for the losses on the line. There is an astounding anecdote concerning repeaters. If a single amplifier was used in the center of a thousand-mile circuit instead of many along the line, the power into the single amplifier would be less than the eye receives from a star too distant to see with the most powerful telescope; on the other hand, the power needed out of this single amplifier would have to be greatly in excess of the total power radiated by all visible stars!

Later, we will look at another kind of repeater -- the type used for *digital* circuits such as T1.

Echos and Singing Look again at Figure 2.4. The symmetrical layout of the drawing suggests a race track. Indeed, unless steps are taken to stop it, a signal will do just what is expected on a race track; it will race around and around. The effect to the observer of this "race" will be a loud howl or screech on the connection. "Singing" is the frequent term used to describe the howling sound.

The thing that must be done is to *balance* the hybrid so that all of the energy arriving at the end of the circuit will be absorbed by that end of the circuit rather than whipping back to the other end. Notice the arrows drawn in each hybrid. Let's examine the "West" end. The energy arriving from the East end is arranged to split so that half goes into the balancing network and the other half into the two-wire section. This is accomplished by attempting to exactly tune the balancing network to look like the two-wire section. Balancing can be quite a technical chore but it can usually be accomplished within reasonable limits.

The energy that originates at West is sent on its way to the East, where similar action occurs. Any imperfection in the balance at either end can cause either a hollow "rainbarrel" effect or singing if the balance is really bad.

*Here's another problem. Telephony existed long before electronics was heard of. There weren't even vacuum tubes, much less transistors and solid-state circuitry. There was, however, another form of amplification that early telephone engineers were aware of. If a telephone receiver is held against a transmitter (microphone) a howl ensues. This means that amplification must be taking place somehow, because one can't have oscillations (howls) without a means that produces more energy out than what comes in to a circuit. The fact is that a carbon transmitter is a device that provides amplification gain. Voila! A device that closely coupled a receiver to a transmitter (so that the sound from the receiver directly actuated the transmitter) was used as an amplifier in the early days of long distance service.

3 PBXs

Some History and Need The earliest telephones were "standalones". Movies made in the 1930s often had scenes of a busy executive, whose importance was apparently judged by the number of phones on his desk.

Not too long after business recognized the value of the phone in conducting its affairs, the need for phones on many employees' desks became apparent. Initially, the interest was in receiving and making calls to customers; however, later the need to call *each other* within the organization was recognized. Thus, the concept of a Private Branch Exchange (PBX) was born.

The original PBXs were, of course, patterned after the telephone central offices which were operator-controlled manual systems.

Manual Switchboard Figure 3.1 illustrates a manual PBX. The principal features are a relatively large number of "jacks" and cords equipped with plugs that fit into the jacks to complete a connection.

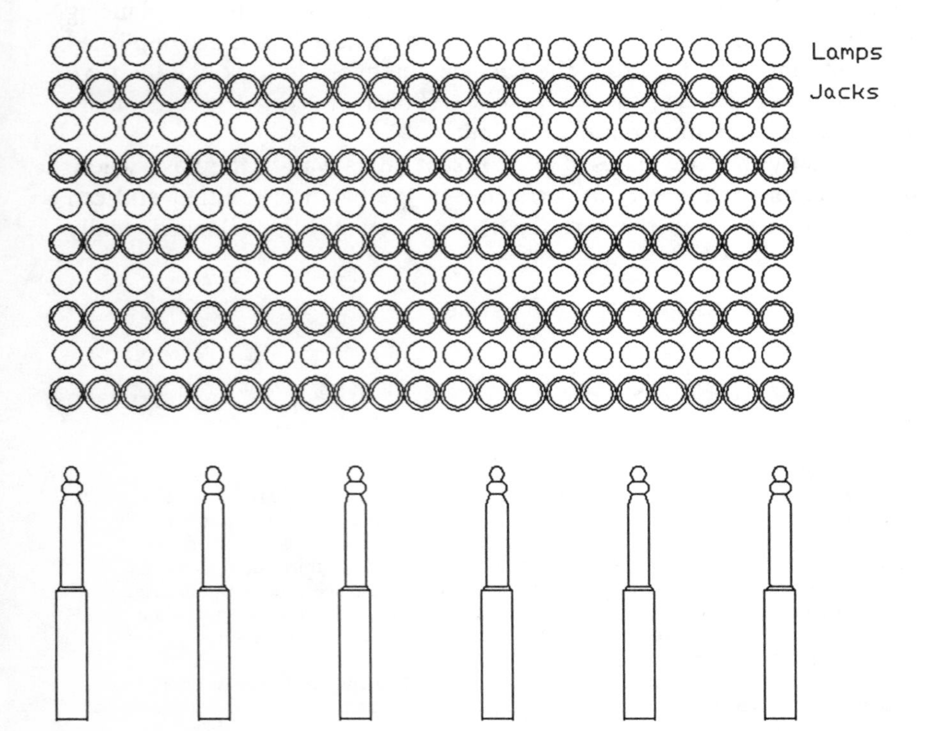

Figure 3.1
Manual PBX

In addition to cords, plugs, and jacks, a means to ring a called *extension* is required. The early boards had a manual crank-operated magneto that produced the necessary ringing voltage. Also, there had to be a way to know when a call was terminated so the connection could be taken down. Again, the early boards used mechanical "drops" that operated when the calling parties "rang off", as they had to do to indicate the end of the conversation. The "drop" system was later replaced by *supervisory lamps* associated with the cords that operated when call termination was detected.

The same drops that indicated the end of a call were used to indicate a request for operator service. These, too, were replaced with lamps in the later boards.

Step-by-Step Earlier we mentioned a man named Strowger, the undertaker. It was claimed of him that as a boy when his mother gave him a task to do, he spent more time figuring out a machine to do it than to do the task itself. Later, when he became an undertaker in Kansas City, he put his inventiveness into play. It is possible that he was a little paranoid since he was convinced that he was losing business to other, similarly-engaged men in town because the operator was giving his calls to them. He set about to devise a system that would let the calling party directly control the establishment of a call; that is, an operator would not be needed in the process. He applied for a patent in 1889 (just 13 short years after the phone was invented by Bell) for what was to become known as step-by-step (SxS) switching. Originally, it was only used for a "central". In this application it was called the "girl-less, out-of-order-less, wait-less telephone". Strowger made his original model out of a collar box, pencil and hat pins. Instead of the familiar dial, the impulses were generated by pushing a number of keys. For example, to call "89", the user had to push one key eight times and the other nine times.

SxS found very wide application as a central office system. Thousands of COs were installed in the 1920s through the early 1950s. Even today, in this electronic age, there are a huge number of COs, particularly in non-Bell locations and England that are in service. However, our interest at this time is in its application as a PBX.

Let's get an insight into its operation. Refer to Figure 3.2. The basis for SxS is a rotary switch, which can distribute an input to ten outputs, depending on its position. Carrying this concept further in Figure 3.3, rotary switches can be connected in a cascaded manner so that the output of one switch can be the input of another. This figure shows how a signal could be sent to 100 locations using two banks of switches connected as shown. Carrying this idea to a PBX, a caller could have access to a total of 100 phones if the switches were correctly set up. The set-up would be in two stages (in a step-by-step manner). Each stage is sequentially set up under the control of the caller's dial. Of course, this concept can be expanded to any desired number of stages to gain access to more and more phones.

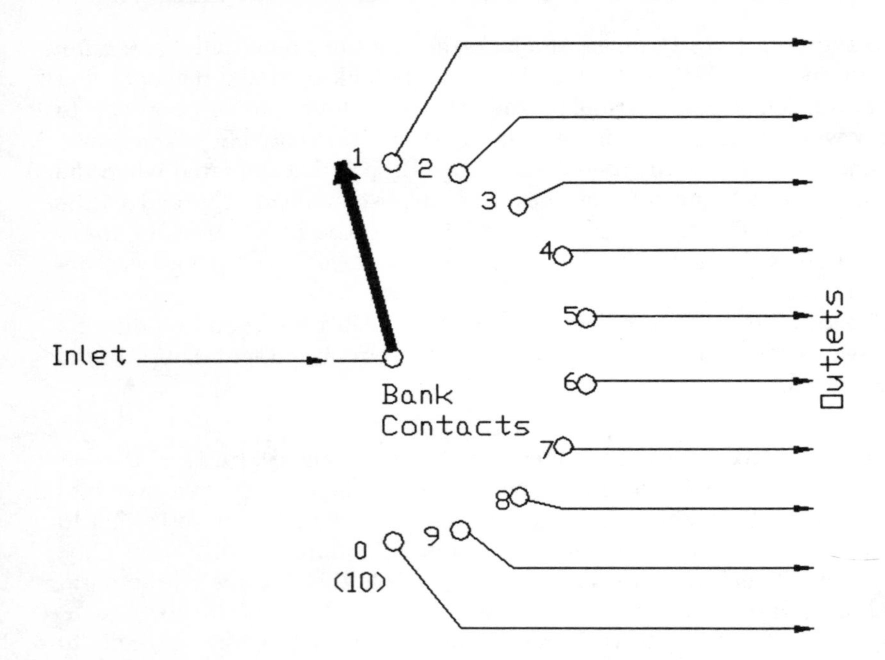

Figure 3.2
Basic Rotary Switch

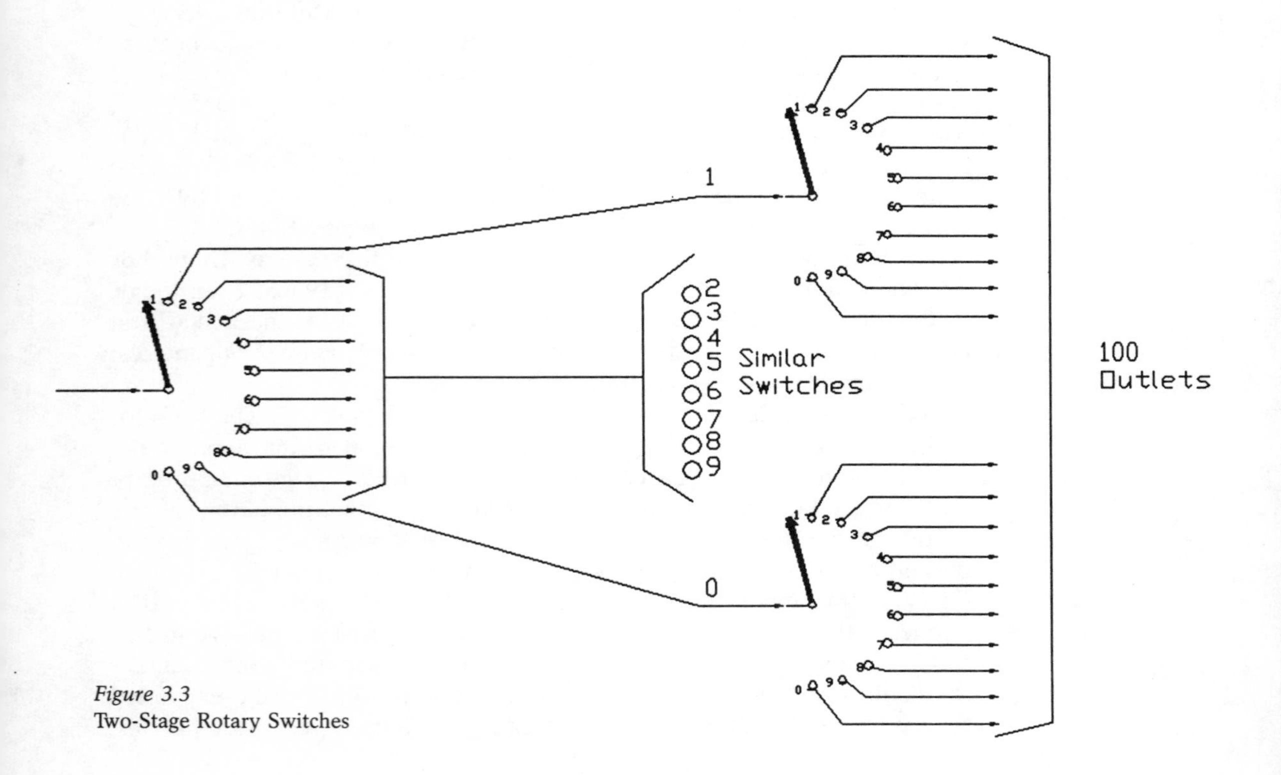

Figure 3.3
Two-Stage Rotary Switches

PBXs, as a practical limit, never exceed five stages, or five digits to be dialed.

Figure 3.4 is a diagram that portrays the usual layout of the SxS switches. This switch arrangement combines two stages of the switches, such as in Figure 3.3, into a single device. It works like this:

The first set of dialed pulses causes the arm contact to go up; e.g., a "5" will raise the arm to the fifth level.

The second set number dialed causes the arm to rotate over to the dialed number; e.g., dialing an 8 causes a rotation to the eighth contact. This results in a connection to "58".

SxS was a marvelous invention for its time.

Crossbar The Swedes were instrumental in the development of telecommunications. The L.M. Ericsson Company, in particular, has been very active through the years in designing innovative equipment. Ericsson designed a fundamental piece of equipment called the crossbar switch.

Figure 3.5 illustrates the principle of this switch. It is composed of vertical and horizontal elements, any combination of which can be connected together to form a path from vertical to horizontal (or vice versa). The connection is formed by electromagnets causing the selected vertical and horizontal contacts to close. Note that every vertical path can be used to connect to every horizontal path. In the simple switch illustrated, a total of four simultaneous connections can be made. Actual switches are made up using ten vertical and ten horizontal elements.

Unlike SxS, the call is *not* set up under the direct control of received dial pulses. See Figure 3.6. A very basic new concept had to be introduced in order to utilize the crossbar switch. This concept is *common control.* Basically, the principle of common control is to gather all of the infor-

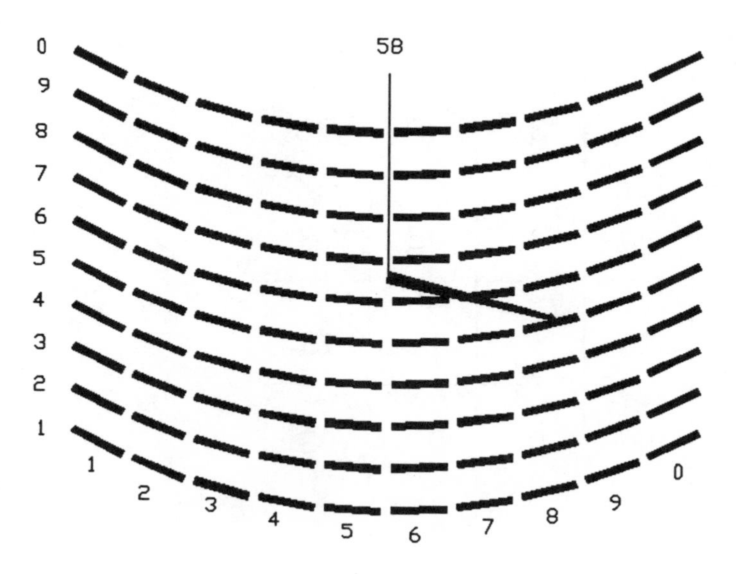

Figure 3.4
Switch Bank

mation needed to process a call and *then* set up the call. Again, this is different from SxS, where the call is set up sequentially as under direct control of the dial pulses.

With common control systems the dialed number is stored as it is dialed. The system determines from examination of the stored digits the proper disposition of the call, setting up the crossbar switches to make the required connection.

The switches are cascaded, one connecting to another to form large systems. The crossbar system has been widely used both for PBXs and central office systems. (The local telephone companies almost completely replaced SxS central office systems with crossbar in the 1950-1960s, that is, until electronic central offices were developed and virtually swept out all of the older offices.)

Trunks and Stations One additional word about *all* PBXs: the central office lines that are provided to access the outside world are called trunks; the local telephones in the system are called stations, or sometimes lines.

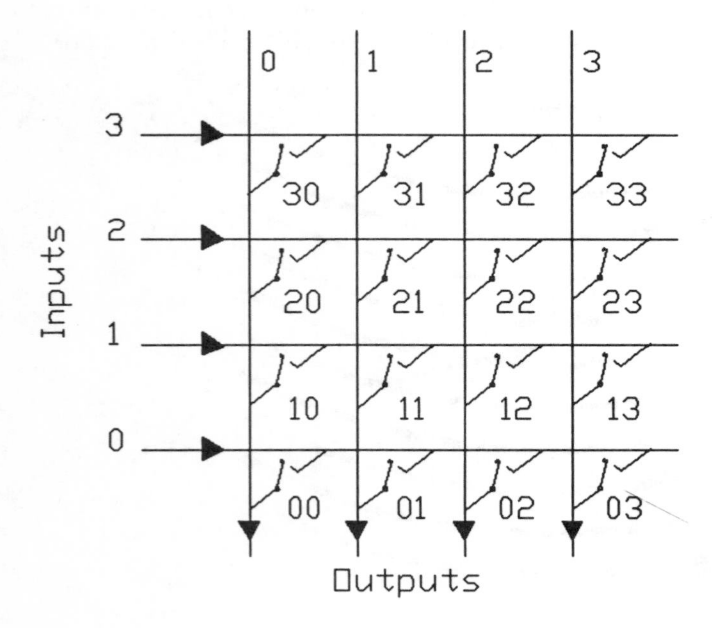

Figure 3.5
Simple Crossbar Switch

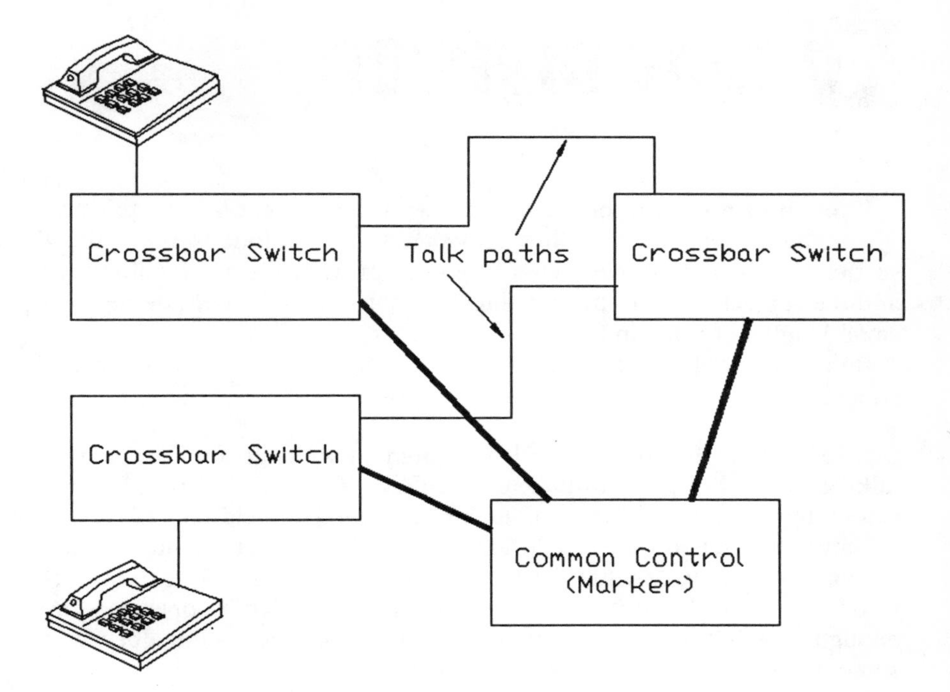

Figure 3.6
Crossbar System

4 TIME DIVISION

Know it or not, everyone is familiar with the concept of time division. We all divide available time into individual slots to accomplish something, so the idea is nothing new. The difference in what we are familiar with and the subject about to be introduced is that we will be considering very small lengths of time, indeed.

Before we look at tiny bits of time, let's look into some fundamental concepts:

Sampling Refer to Figure 4.1. Imagine that we have a telephone user talking to another person by means of the wire shown. Notice that there is a knife switch in the circuit that can be opened and closed.

Naturally enough, one would expect that the listener would hear lots of clicks and some of what the talker is saying. This is true if the switch is opened and closed slowly; however, if the switch is operated fast enough, the listener would begin to hear all of what was said, albeit with some noise. But, how fast is fast enough?

It turns out that fast enough is at a rate twice as high as the highest frequency that is to be heard. In the case of speech, this is 2 × 4000, or 8000 times per second. (Good quality speech requires a top frequency of 4000 hertz.) It is interesting to note that it does not matter how long the switch is closed when it is closed for the sample to be useful in carrying voice information; just so it closes at least 8000 times per second.*

Figure 4.2 shows graphically the result of doing this to a typical voice signal. Notice that the outline, or "envelope" of a portion of the voice signal we saw in Figure 2.2 is still apparent in Figure 4.2; however, it is chopped up into individual pulses.

Instead of one knife switch as in Figure 4.1, let's add another and put each of them at the respective end of the wire, as in Figure 4.3. If we carefully arrange it so that each switch is opening and closing in perfect synchronism with the other we can expect that the system will still work. We have a reason for doing this, as we will see next.

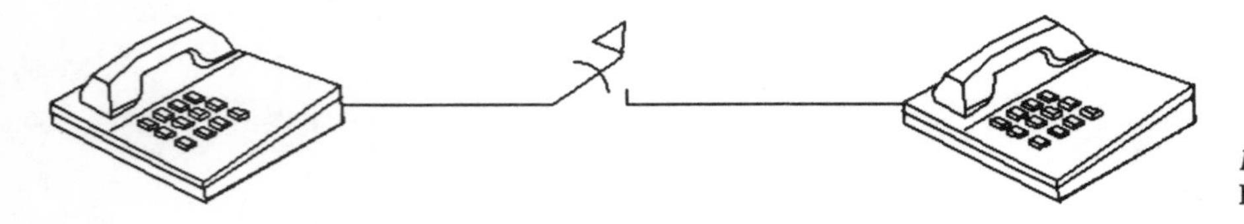

Figure 4.1
Basic PAM Concept

*As a matter of interest, digital compact disks that provide very high fidelity music recordings use a sampling rate of 44000 times per second to obtain frequencies up to 22000 hertz.

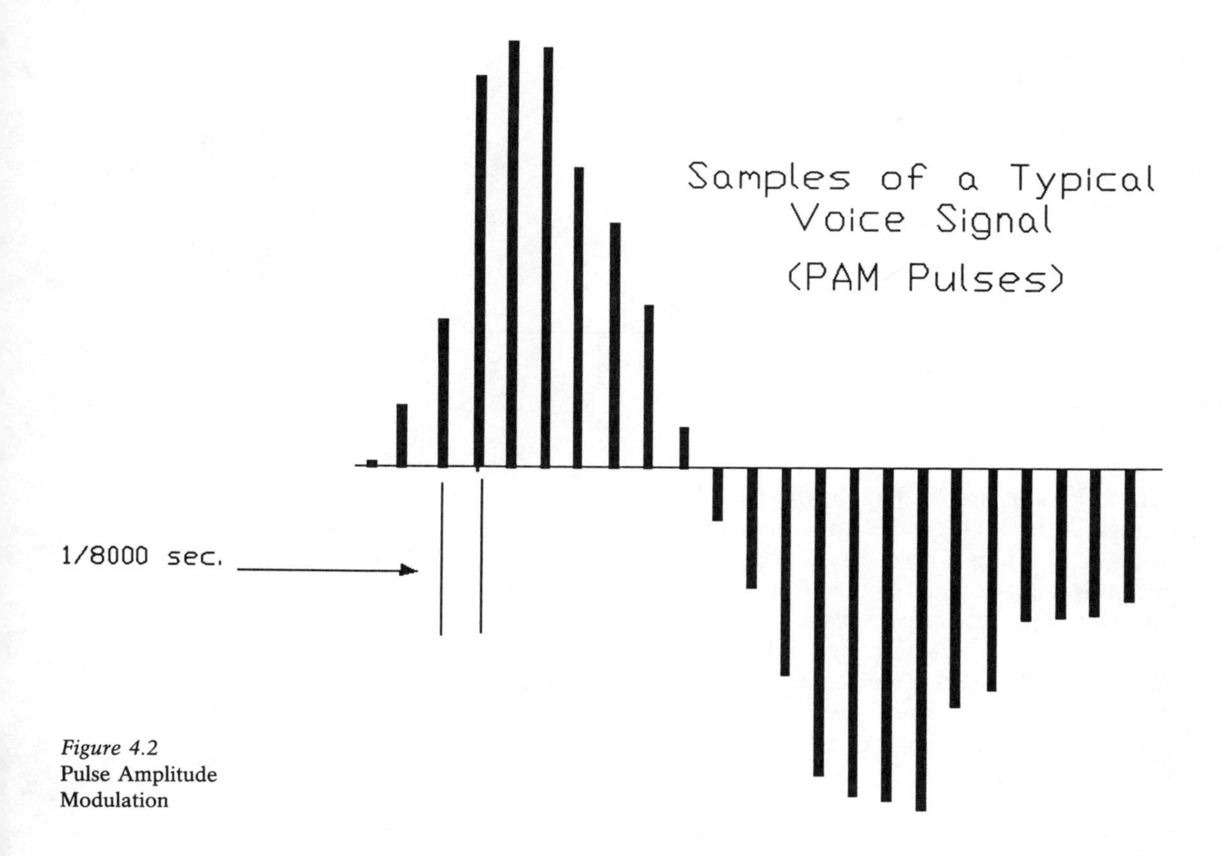

Figure 4.2
Pulse Amplitude Modulation

Refer to Figure 4.4. Instead of simple knife switches, we see something that closely resembles two car distributors connected together by a wire. The two devices are actually called commutators. They are arranged so that they rotate in perfect synchronism; when the *rotor* is connected to segment 1 on the left commutator, it is connected to segment 1 on the right commutator. Furthermore, the rotors are rotating at 8000 revolutions per second. A moment's reflection will convince the reader that this is a very high speed (480,000 RPM). Obviously, the commutator cannot be a mechanical device, but must be electronic. Also, if the wire is very long, there must be an accounting for the *propagation delay* of the samples and the receiving end's commutator must be suitably delayed.

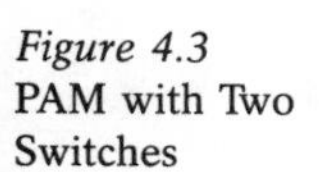

Figure 4.3
PAM with Two Switches

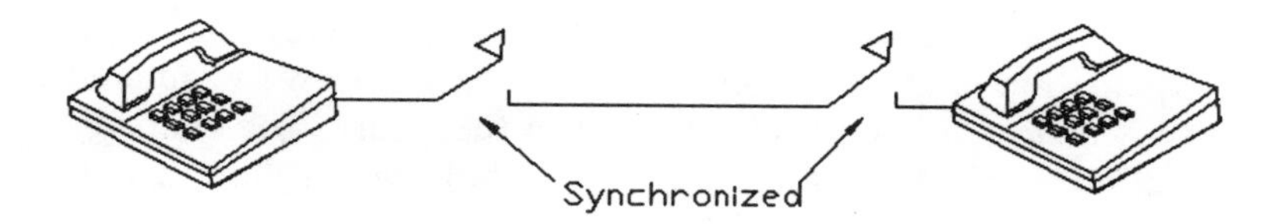

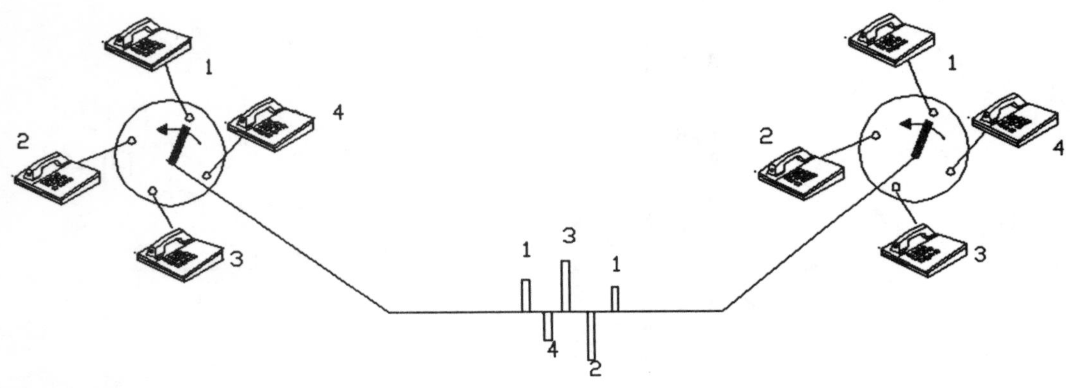

Figure 4.4 Four Simultaneous PAM Calls

Let's look at practical applications of this system. A telephone is shown connected to each of the four segments, 1, 2, 3 and 4. It is somewhat intuitive that four talkers can now carry on a conversation as long as each is connected by means of the electronic commutators 8000 times per second and in perfect synchronism.

The technique just described, where signals are sampled is called *Pulse Amplitude Modulation,* or PAM. Nowadays, it is used only as a first step in time division or *digital* methods. It is a very important concept to understand since it forms the first building block in a digital switching or transmission system.

Digitizing As useful as PAM is, it has some major disadvantages such as vulnerability to "noise" such as pops and clicks on a telephone line. Furthermore, today's electronic components are designed to inexpensively handle signals that are simply ON and OFF, or *binary* signals. Stated another way, the chips so commonly used today cannot accommodate signals of varying amplitude, such as the PAM pulses we just discussed.

A means is needed to convert the PAM pulses into simple ON and OFF (binary). This conversion process is called *digitizing.* First, it must be recognized that the PAM pulses can have virtually an infinite number of possible levels, or amplitudes. We are going to artificially and arbitrarily assign a value of amplitude that is "close" to the actual value. Obviously, the more values we permit for comparison, the closer to the true amplitude we will be.

Pulse Code Modulation - PCM Looking at Figure 4.5 we see that we are arbitrarily permitting only eight possible values; this is to make the explanation easier to envision. The PAM pulses we examined earlier are shown *superimposed over* a chart in the figure. The chart divides the amplitudes into eight ranges. If a pulse falls within the range, it

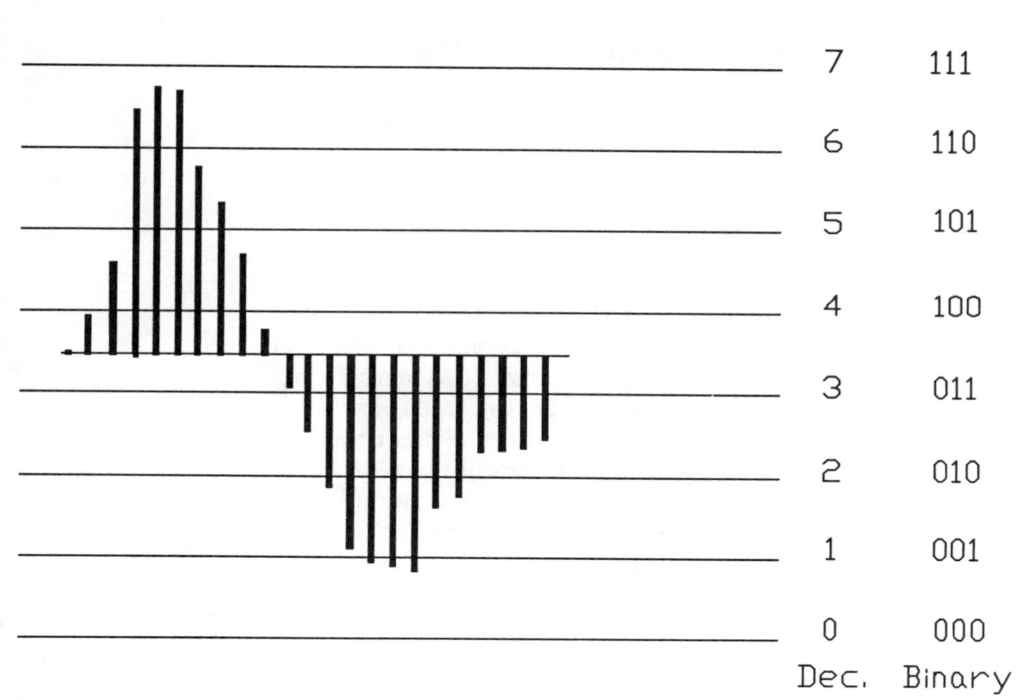

Figure 4.5 Conversion of PAM to PCM

is considered to have that area's amplitude. The operation of assigning approximate values is called *quantizing*. Now, here is the important part. Notice on the left of the chart that there is a series of numbers that mathematically represent the ranges. The familiar decimal numbers 0-7 and the not-so-familiar binary numbers of 000 to 111 are shown. In summary, what has been done is to examine each PAM pulse and assign a numerical value to it. The whole process just described is called *Pulse Code Modulation,* or PCM.

Now that a numerical value has been assigned to the PAM pulses (which represent the actual signal), these numbers can be transmitted from the talker to the listener's end of a circuit instead of the PAM pulses. Why would we want to do this? There are a number of excellent reasons we will examine shortly, but first let's smooth out a few rough edges in the discussion.

It was alluded that the more quantizing levels there were the closer we could come to the original PAM pulse amplitudes. This is certainly true, and indeed the use of only eight levels, as in the figure, results in rather poor reproduction of the original sound. However, the more levels that are used, the larger the number of values that must be transmitted to represent the signals. In actuality, commercial systems use 256 levels; this results in excellent sound reproduction. It happens that binary numbers in the range of 00000000-11111111 can represent the 256 levels, ($2^8 = 256$).

Eight bits are needed to represent the levels. Having this knowledge, we can perform a little arithmetic:

8000 samples per second
8 bits per sample

$8000 \times 8 = 64{,}000$ bits/second

In other words, good telephonic reproduction requires the transmission of 64,000 bits/second (64 kilobits/sec.). So far, it seems that we have made a poor trade. A voice signal whose highest frequency is 4000 cycles per second requires 64 kilobits per second for transmission. Again, we will shortly look at the numerous advantages to making this trade.

Referring again to Figure 4.4, the one with the commutator, we saw that four talkers could share a single circuit by means of the PAM methodology. Again, commercial systems are designed to carry 24, not 4, simultaneous conversations. The 24 sets of PAM pulses are quantized into 256 levels each. Looking again at the arithmetic:

64,000 bits per talker
24 talkers

$64{,}000 \times 24 = 1{,}536{,}000$ bits per second

Refer to Figure 4.6. The terminal of a North American commercial PCM transmission system called *T1* is illustrated. From the above, we see that to handle 24 conversations digitally, requires the transmission of over 1.5 *megabits* per second. In addition, "housekeeping" bits for timing and synchronization are added to bring the total to 1.544 megabits per second.

Figure 4.6
Voice Channel Terminal for T1

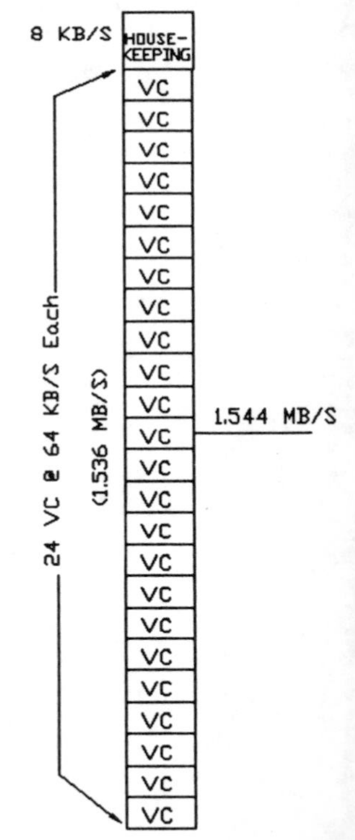

Regenerative Repeaters The subject of repeaters was introduced in Chapter 2, where it was noted that repeaters were first used to "repeat" telegraphic pulses. It was also noted that the repeaters were amplifiers used to strengthen weak analog voice signals. However, repeaters for digital signals are not simply amplifiers that strengthen digital pulses; as a matter of fact, they closely resemble the early telegraphic repeaters. Repeaters for digital signals are called "regenerative repeaters" since they *create* brand new pulses to send along the line.

Figure 4.7 illustrates what happens. Pulses transmitted along a telephone line soon become more like smears than pulses. This is due to the fact that an ordinary telephone line was not really designed to carry pulses, but to carry analog voice signals.

Figure 4.8 shows how smeared pulses enter a regenerative repeater, and this is the important part, the repeaters examine each incoming smeared pulse and decide whether the pulse is a binary one or zero. Based on the

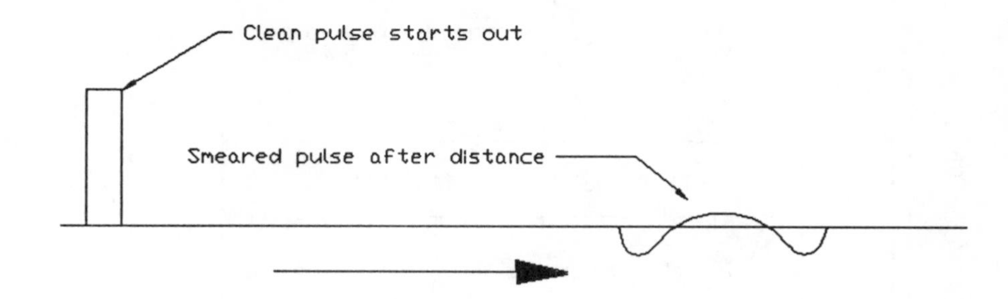

Figure 4.7
Effect of Line on Pulses

decision made, either a brand new one or zero pulse is created and sent along the line. Noise or other impairments are not amplified and also sent along. Hence, if the repeaters are spaced close enough together to reliably determine whether a pulse is a one or zero, near perfect transmission fidelity can be anticipated. This certainly is not true with analog systems where noise and other impairments are strengthened and sent along from repeater to repeater.

Figure 4.9 (next page) is a diagram of an overall T1 system, showing the regenerative repeaters in each direction of transmission. The repeaters are spaced (usually about 6000 feet) close enough to be able to reliably regenerate the 1.544 mb/s pulses.

Conclusions Digital techniques are used in transmission systems and switching systems, including PBXs. The reason for the virtual explosion of digital systems can be most directly laid to the advent of microprocessor chips. These chips all have one thing in common: they are cheap, and they are binary (two-state) devices. Since unmodified voice signals are

Figure 4.8 T1 System Regenerative Repeaters

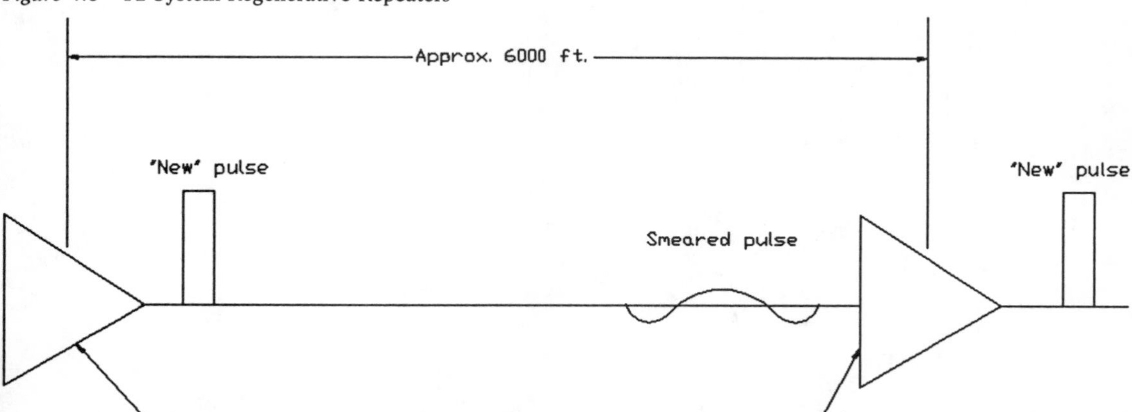

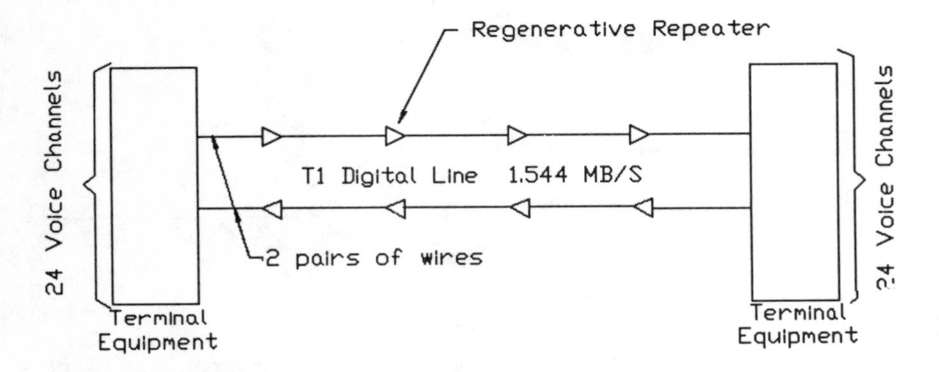

Figure 4.9
T1 Carrier System

not digital, but smoothly varying (analog) in nature, they must be modified before advantage can be taken of the very inexpensive modern electronics and this is what digitization by PCM does. If digital techniques weren't available, telephony would be left behind in the computer revolution. Instead, telephony is in the mainstream of this exciting arena.

5 PBXs II

Enter Electronics Of course, electronics *per se* are not a recent development. In fact, the invention of the vacuum tube by Lee DeForest in the 1920s ushered in the "electronic age". It wasn't until World War II, however, that electronics really came to the forefront as a science. Radar, sonar and gun control devices developed mainly in the United States were truly sophisticated devices that applied principles that are still greatly in use today. Nevertheless, it still took the invention of *solid state* devices such as diodes and transistors and, most importantly, the so- called *chips* to skyrocket electronic usage.

First Electronic PBXs Early attempts were made to utilize tubes, and then transistors as switching and logic devices in place of electromechanical relays and step-by-step switches in PBXs. No serious attempt was made to actually replace the step-by-step *per se,* although one could envision such an arrangement. Crossbar systems, however, were a "natural" for electronic implementation since they used common control logic. Indeed, the earliest systems retained crossbar switches for the "talk path". The initial effort was in using electronic logic circuits in place of relay logic.

The reason most efforts were directed toward the use of electronics for only the logic portion of systems was because at the time metal-to-metal contacts were the only reliable and efficient talk path crosspoints available. (For the record: the very first electronic *central office* used especially designed neon gas tubes for crosspoints.)

While all of the early developments of electronic PBXs were taking place, the concept of *stored programming* was being developed. It was this effort that not only heralded the development of computers as we know them, but true electronic PBXs. Before we go deeper into this topic, we need to establish some more principles. The topics of space and time division, which will be covered next, relate to the switching network employed in a system, in other words, the means that are used to establish the talk path.

Space Division The concept of space division is so obvious that it is difficult to describe. What is meant is that *physical space* taken up by the network is in proportion to the size of the network. (Not a very profound statement, but it is the essence of what the author is attempting to put into simple terms.) The bigger the PBX (or Central Office) is, the more space is needed for the network. Examples of space division networks are SxS and crossbar, already discussed. On the other hand, time division

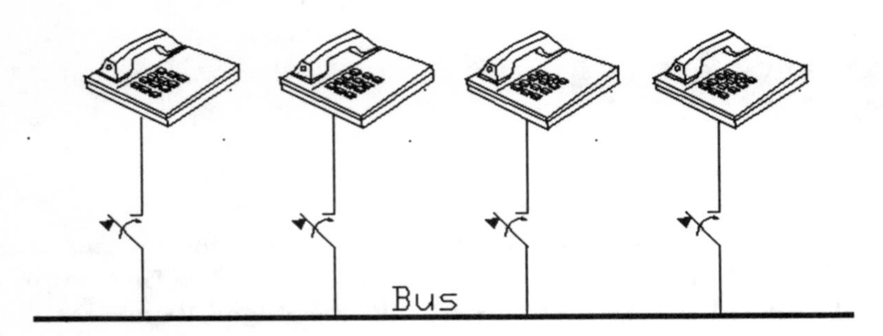

Figure 5.1
PAM PBX Principle

does not really demand more and more space as the system gets larger. (Not *strictly* true, but it is useful to think of it this way since the required physical space increases only in increments.)

Time Division The topic of time division was introduced earlier when we examined PAM and PCM techniques. Precisely the same techniques can be applied to switching. First, let's look at the PAM technique as applied to switching. Refer to Figure 5.1. In this figure we see four phones connected by knife switches to what is called a "bus". This arrangement is a little different than the commutators we looked at earlier since the switches don't have to open and close in a fixed sequence. In fact, *pairs* of them are operated in a definite time sequence to connect phone pairs together for conversation. (Bear in mind that the switch closures are occurring at an 8000 times per second rate). Each pair is assigned a *time slot* for operation. See Figure 5.2.

As noted above, this is the Pulse Amplitude Modulation (PAM) method of operation. There are a number of successful PBXs that have used this very method to provide the switching matrix. The most well-known is the Dimension® series of PBXs. The primary drawbacks are the very ones mentioned earlier. They are not immune to noise and cannot utilize the inexpensive binary (ON/OFF) devices such as chips. Also, an increasingly important consideration for PBXs—the matrix cannot carry high-speed digital data without the addition of modems. This topic will be covered in detail later.

The solution to these shortcomings is the use of PCM, where the PAM pulses are digitized in the same manner as discussed earlier.

This concept is entirely analogous to the case where we used PCM to

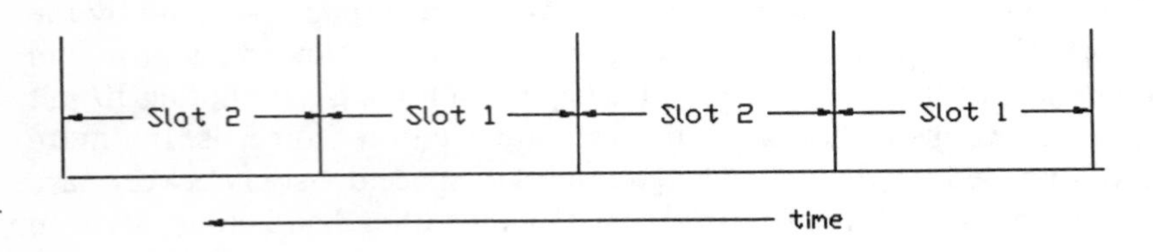

Figure 5.2
Diagram of Time Slots

provide transmission along a wire, as discussed in Chapter 4. In this case, however, we will use PCM to provide switching paths between phones. The simple switches in Figure 5.1 will be replaced by "black boxes" that convert the analog voice signals to binary (digital) signals. Recalling the arithmetic of the earlier discussion:

8000 samples per second
8 bits per sample

$8000 \times 8 = 64{,}000$ bits/second

Refer to Figure 5.3. Each voice signal is converted by the black boxes (actually called "codecs") to 64,000 ON/OFF pulses per second. It is these pulses, or bits, that are carried on the bus. And now that they are truly digital, they can be handled by the inexpensive chips and also can be manipulated exactly in the same way as computer data. Again, digital data signals from users' terminals, as we will see later, can be carried without the use of modems.

Software Control Other than SxS, earlier PBXs used what is called, "wired-logic" to control their actions; e.g., providing dial tone, registering the dialed digits, call routing and setup, etc. Every possible action had to be anticipated and the circuitry designed and wired to perform the action. For the most part, electromechanical relays were utilized to perform the logical functions, or more simply, "logic". The early electronic PBXs followed this pattern, but used solid-state devices to perform the logic. In the meantime, though, electronic central offices were using stored program software control.

Stored program software control, or "software", is quite similar to programming a computer. Again, every possible required activity or anomaly that might be encountered has to be anticipated and covered by the software logic.

Referring to Figure 5.4 (next page), we see that the software-controlled central processor CPU has a number of functions to perform:

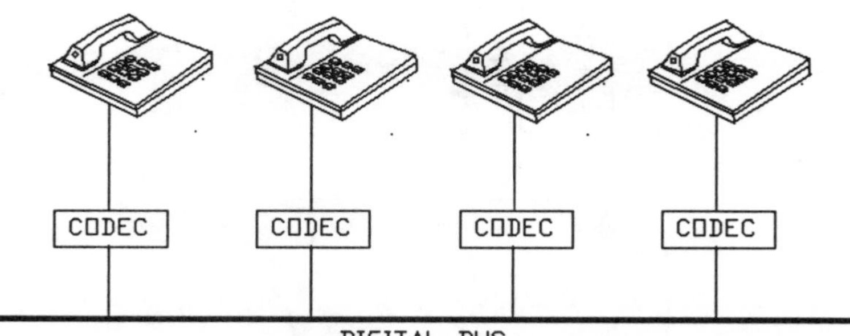

Figure 5.3
PCM PBX Principle

Scanning of lines and trunks	Call setup
Digit registration	Ringing and signaling
Routing of calls	Call accounting

Each of these actions must be programmed in detail for the CPU to perform its functions. In addition, careful attention must be paid to timing; for example, all of the dialed digits must be "caught on the fly" from all users that happen to be dialing at that moment. Of course, with the exceedingly high speeds of computer chips, this is not difficult with proper design.

PBX Features The advent of software-controlled PBXs brought with it a host of new features, since additional features usually mean only the addition of new software, not "hardware", as with the old PBXs. Some of the features are esoteric and not too beneficial to most users. The following list is a sampling of what can be found on modern PBXs without specialized telephone sets, although many systems offer the option of esoteric executive sets:

Consultation hold
Three-way conversations

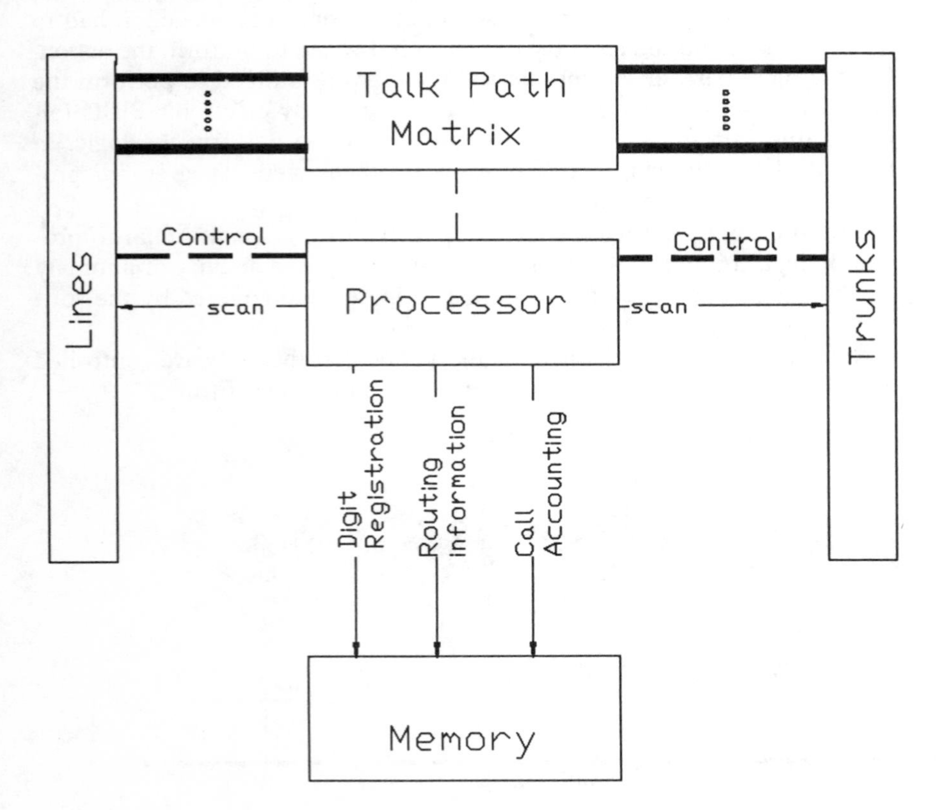

Figure 5.4
Electronic PBX Block Diagram

User transfer
Paging from extensions
Telephone dictation
Terminal hunting
Call forwarding
Call waiting
Automatic call back
Barge-in
Speed calling
Toll restriction for certain stations
Queuing for outgoing lines

Tie Lines There are some features that are common to most PBXs, and, in many cases, they have been in existence for a number of years. An example of this is tie lines. A tie line is simply an arrangement that provides a "permanent" circuit between two or more PBXs, as shown in Figure 5.5. Usually, the circuit is leased from a common carrier; however, they can be provided by user-owned facilities such as private microwave. A tie line is very similar to a trunk. Its purpose is to provide direct access from PBX to PBX.

In its most common configuration it is arranged so that dialed information is sent along to select a station at the distant PBX. Access at the calling PBX is made by dialing a digit such as "8". At this time, it is common for the caller to hear dial tone from the called PBX. However, with electronic PBXs, access may be made without hearing the distant PBX dial tone; rather, the user simply dials the desired 3-4 digit extension number following the access code.

Figure 5.5
Simple Dial Tandem

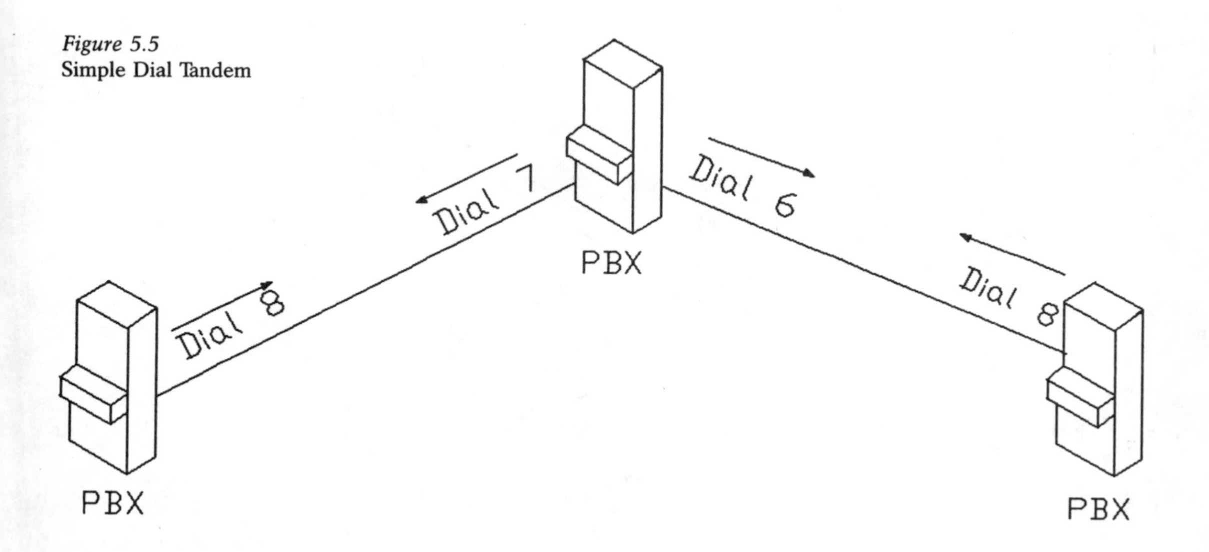

Later, we will examine the methods used to provide signaling between the PBXs.

Tandems Large firms with many corporate locations are likely to have PBXs in most of their areas. Of course, carrier-supplied dial up service can be utilized to call between locations; however, this sometimes is not as cost effective as the use of tie lines between locations. Again, it usually would not be feasible to provide tie lines between all locations, such as in Figure 1.5. What is needed is a central switching location at the geographical center of the user complex, as illustrated in Figure 5.5.

The figure shows that one of the PBXs has been assigned a non- PBX-like function—tandeming. The tandem switch serves to establish tie line to tie line connections. The switch is "told" by the incoming tie line to connect to a particular group of outgoing tie lines, usually by a single "directing digit" such as "7". Larger systems might require two or more digits. Usually, all of the tie lines are able to serve both as incoming and outgoing facilities. Also, the PBX can originate or receive calls over the same tie lines.

Stored-program electronic PBXs have changed the concept somewhat. Refer to Figure 5.6. In this example, the centrally located PBX is "smart" enough to know by the digits dialed into it from the remote PBX where to connect the tie line. The extension number is then outpulsed to the desired PBX. Now the overall complex of PBXs have become integrated into what is called a *Network*.

Network Refer to Figure 5.7. Fundamentally, a network is the same as tandeming that was described earlier; however, a universal numbering

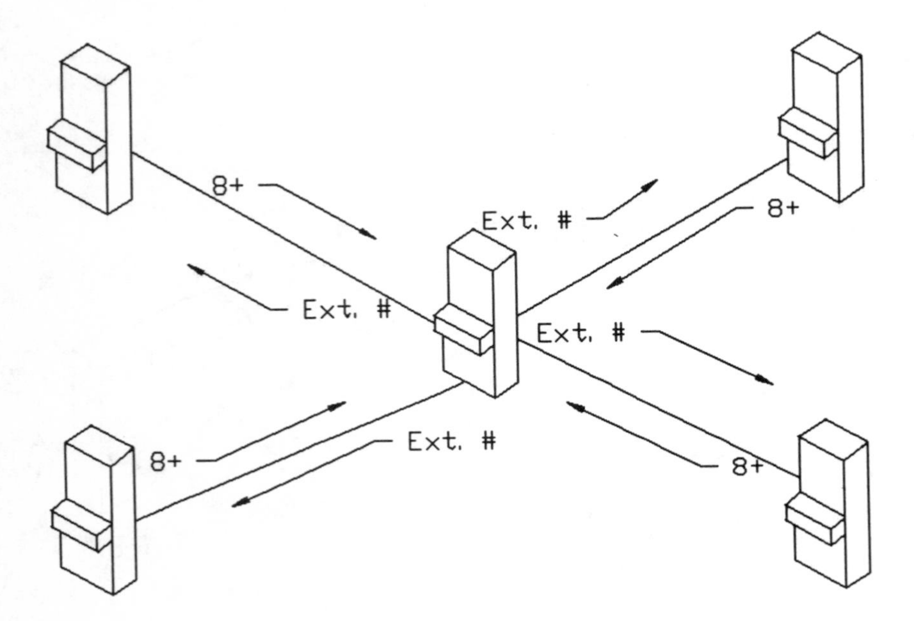

Figure 5.6
Electronic Tandem

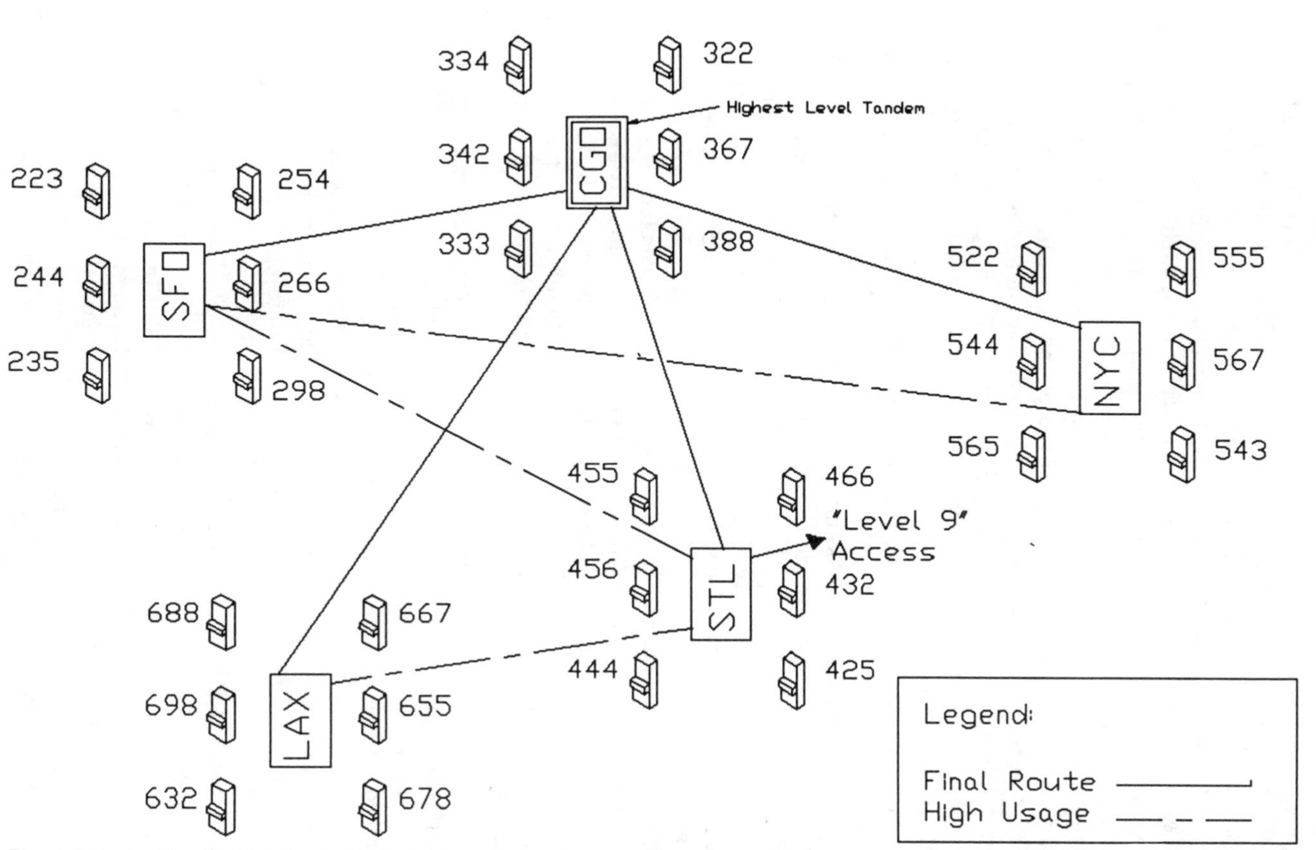

Figure 5.7 Sophisticated Electronic Tandem Network

plan is established overall. Each PBX homes on an electronic tandem switch (labled as CGO, SFO, NYC, LAX and STL) and all of them are given three-digit codes, such as 223. Each telephone in the system is given a unique four-digit number such as 5634, corresponding at least in part to the extension number; e.g., 634. Generically then, each phone is identified by a code, NNX-XXXX. Further, the user does not have to supply "directing digits" as was true with simple tandem systems. In the usual case, the user wishing to call another location only has to dial "8" and follow it by the seven-digit number of the phone he wishes to reach.

In this figure we see that each PBX serving as a tandem is "intelligent" enough to decode the NNX digits to determine the proper group of tie lines to route the call to. A fairly large network is illustrated in Figure 5.7. Note that there can be more than one tandeming point in the network, and that more than one possible route is available. This is called *alternate routing.*

Let's look at an example. Assume a station user in the PBX labled 567 that homes on NYC wishes to call the user in the San Francisco area identified as 223-5634. First, he dials "8" into his PBX, which connects him to the tandem switch his PBX homes on -- the NYC switch. Upon receiving dial tone from the NYC switch, he dials 223-5634. Note that there are *high usage* circuits shown that directly connect NYC to the SFO switch. If any of these circuits is available, it will be selected. The digits, 223-5634 would be sent to the SFO switch directly. If none is available, the NYC tandem "knows" that call should be sent to SFO via the *final route* circuits that go to CGO. The CGO switch then sends 223-5634 on to SFO. The SFO switch sends only 5634 to the PBX, where the call is completed.

Calls to the *public network,* can be sent over the private network to "hop off" at a point that provides the least cost. Notice that a "level 9" circuit is shown on the STL switch. All of the tandem points and/or PBXs can provide 9th level access to the public network.

Many pages could be devoted to networking. This compressed discussion was only devoted to a portion of the considerations for a *voice-only* network. Many more pages could be devoted to data applications.

Direct Inward Dialing - DID In the previous discussion we saw where digits were sent directly to the PBX to select the called phone. Of course, this was with a tandem network. What about the case where calls are made to a PBX from the public network through the local central office? It is quite useful to bypass the PBX attendant if the caller knows the extension number he wants. This service has been available for years and is called Direct Inward Dialing (DID).

Refer to Figure 5.8. The serving telephone company reserves a group of telephone numbers for a large PBX. When one of these numbers is called, instead of ringing the PBX on a trunk, the CO pulses the digits of the called extension into the trunk. Either dial pulses or tone pulses (DTMF) can be utilized.

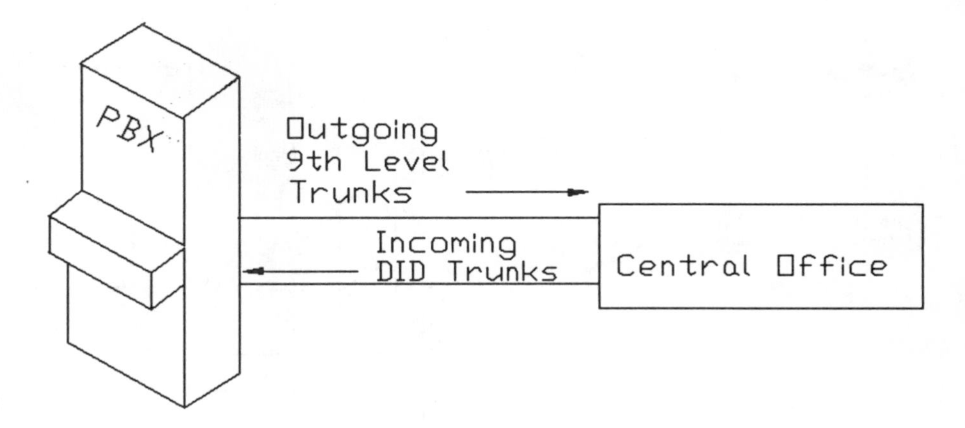

Figure 5.8
Direct Inward Dialing

There are some recent developments that permit calling of the desired extension directly by the caller by means of his tone dialing pad; in this way a valuable group of numbers does not have to be reserved by the telephone company.

Centrex Centrex can be thought of as a replacement for PBX service. In simple terms Centrex provides PBX-like service directly from a telco central office without any major equipment at the users' premises. A pair of wires for each user's phone is extended from the CO to the user's office location. See Figure 5.9 (next page).

A moment's reflection will give a person reason to question the economy of this. It would certainly seem that this is a waste of copper wire and inherently expensive. Certainly, one would think, a PBX on the premises acts as a concentrator—many phones can be served by relatively few CO trunks. This is true; however, telephone cable is not expensive and there are a number of fine advantages in having service provided directly by a central office:

Inherent direct inward dialing
24-hour maintenance coverage
Excellent backup against power outage
Skilled personnel in attendance
Economies due to large scale
Large redundancy of common equipment
Easy upgrade to new features
No capital investment by user for PBX
No need for large floor space allocation

Call Accounting Another feature of Centrex *and* modern PBXs is a means to post and even price out toll calls as a way of accounting. Of course, call accounting has been a feature of telco service for about as long as it has existed. It is only in the past ten years or so that users were

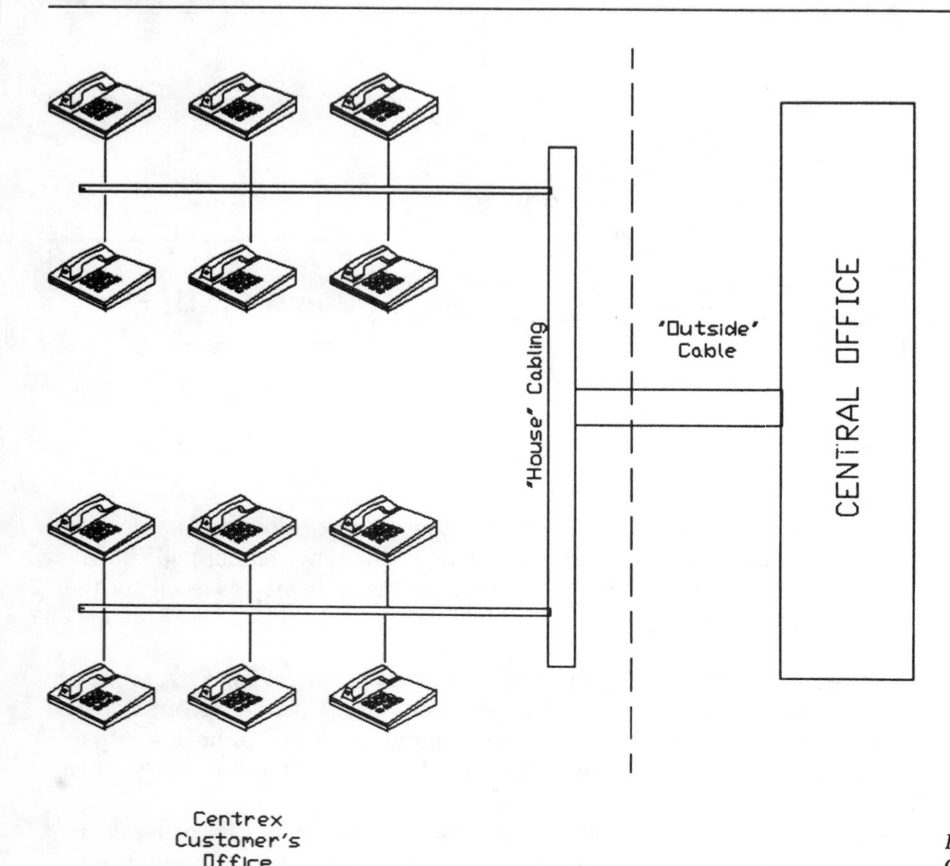

Figure 5.9
Centrex System Layout

able to reconcile their own phone bills and allocate costs to various departments. Since Centrex is a central office service, call accounting is "free". However, PBXs must have it added as a software/hardware feature. See Figure 5.10.

The feature goes by a number of names, including most commonly SMDR (Station Message Detail Recording) or CDR (Call Detail Recording).

Call detail information is stored in the PBX until all of the call information, including duration, is available. The information is transferred via a "standard" RS-232C interface to be processed.

In almost every case, the processing equipment is provided as an adjunct piece of apparatus, with the PBX supplying the call details such as:

Date
Time
Duration
Station number (extension)

Trunk route
Dialed number
Cost

The SMDR/CDR equipment primarily consists of a specialized computer that has storage capability and a database of rating tables that permit reasonably accurate pricing for internal call reconciliation.

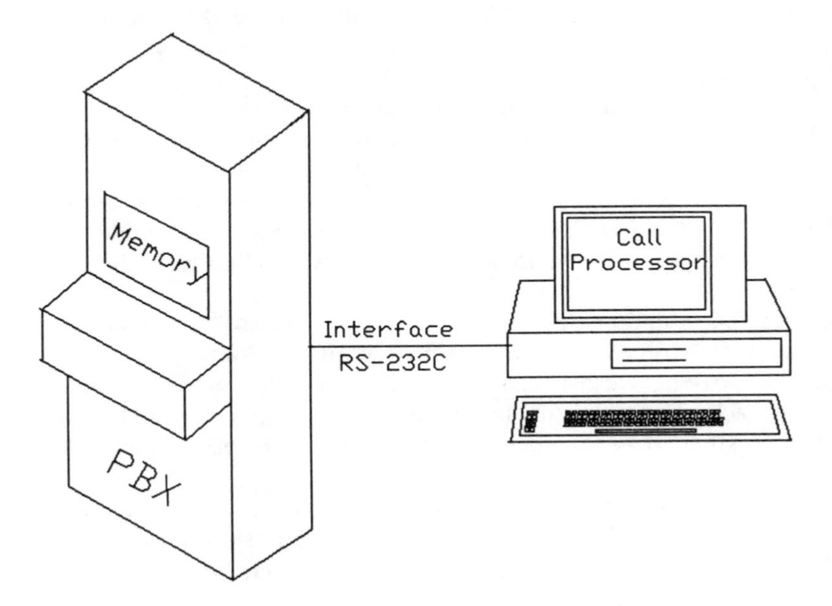

Figure 5.10
Call Detail Recording

6 KEY EQUIPMENT

Background One measure of a businessman's success, as depicted by Hollywood years ago, was the number of telephones he had. It wasn't uncommon for an executive to have three or more of the old "candle-stick" phones taking up space on his desk. An illustration of this is shown in Figure 6.1. Clearly, this was a situation that called for some inventiveness. The local telephone companies came to the rescue with some equipment that resembled a glorified cigar box equipped with some "keys" or switches to enable the user to have access to a number of lines with a single phone and be able to place calls on "hold".

For a number of years each telephone company (and even areas within a local telephone company) developed its own circuitry to provide this service. The result was a hodgepodge of unique systems that were difficult to maintain.

To resolve this situation Bell Labs developed "standard" arrangements which eventually evolved into the extremely popular 1A2 Key Telephone System. This system, which is characterized by the "fat" wires or cables attached to each phone and the set of blinking lamps, is to this day the most likely system to be found in an office. However, electronics have made their mark here, too.

A large number of manufacturers have developed electronic systems that eliminate the fat cables, in fact, they are called "skinny" wire systems. In addition, as we shall see, they offer a myriad of features not found on 1A2 key systems.

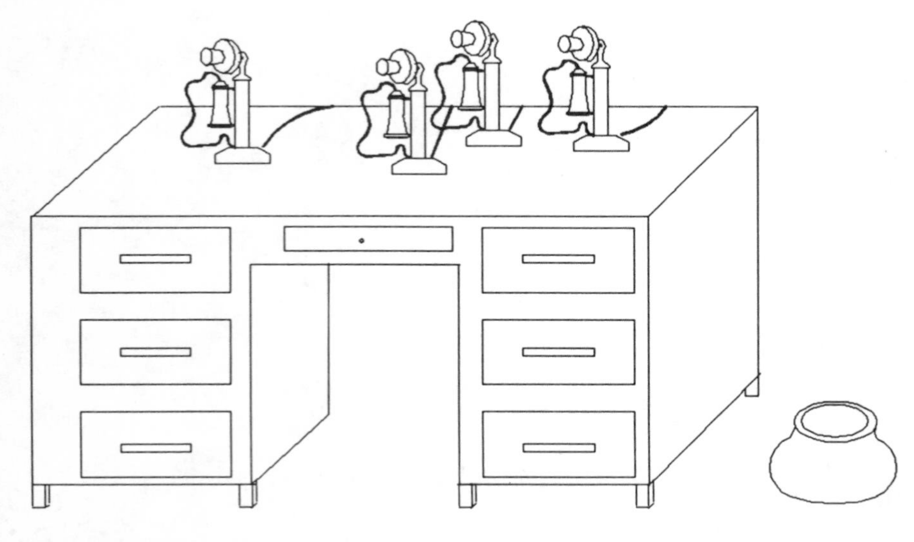

Figure 6.1
Busy Executive's Desk—1930

Fundamentals It is important to note the difference between modern key equipment and a PBX system since they can greatly resemble each other. The following chart outlines the more common features and the *usual* differences:

FEATURE COMPARISONS

	PBX	**KEY**
Telephone sets	Simple	Complex
Access to CO lines	Dial 9	Direct selection
Answer incoming calls	Attendant console	Any user answers
Size of system	Generally large	Usually less than 20 phones
Calling emphasis	Between office users	Outside calls
Intercom	Many paths	Few paths

The economics and practicality of using a PBX instead of key equipment for, say, 20 phones or less has to be carefully examined.

Station Equipment The most obvious thing to note is that the telephones used in key systems are usually sophisticated devices having many buttons or "keys" that are primarily used to access a number of lines. These enable the user to either place or receive calls on any of these lines. See a typical 1A2 6-button set in Figure 6.2.

Figure 6.3 shows equipment that was developed because of the need for a centralized attendant to answer a large number of lines. This re-

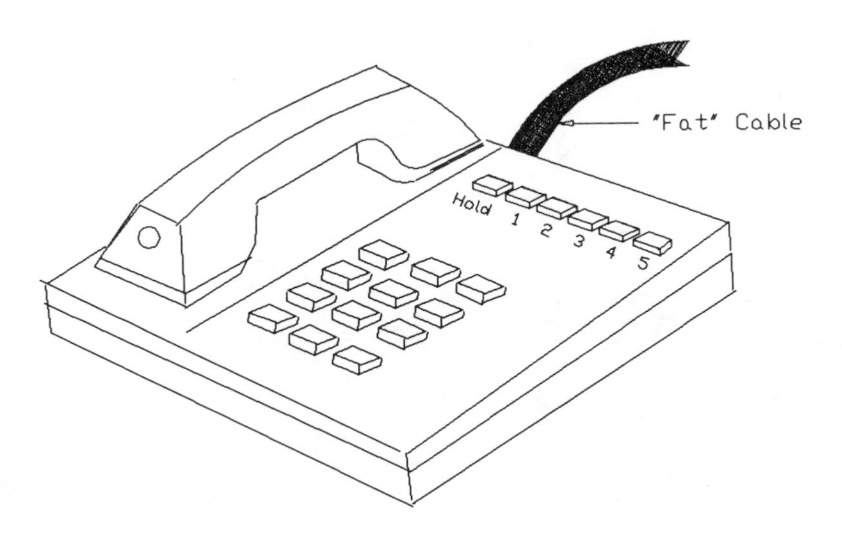

Figure 6.2
Older-Type Key Telephone

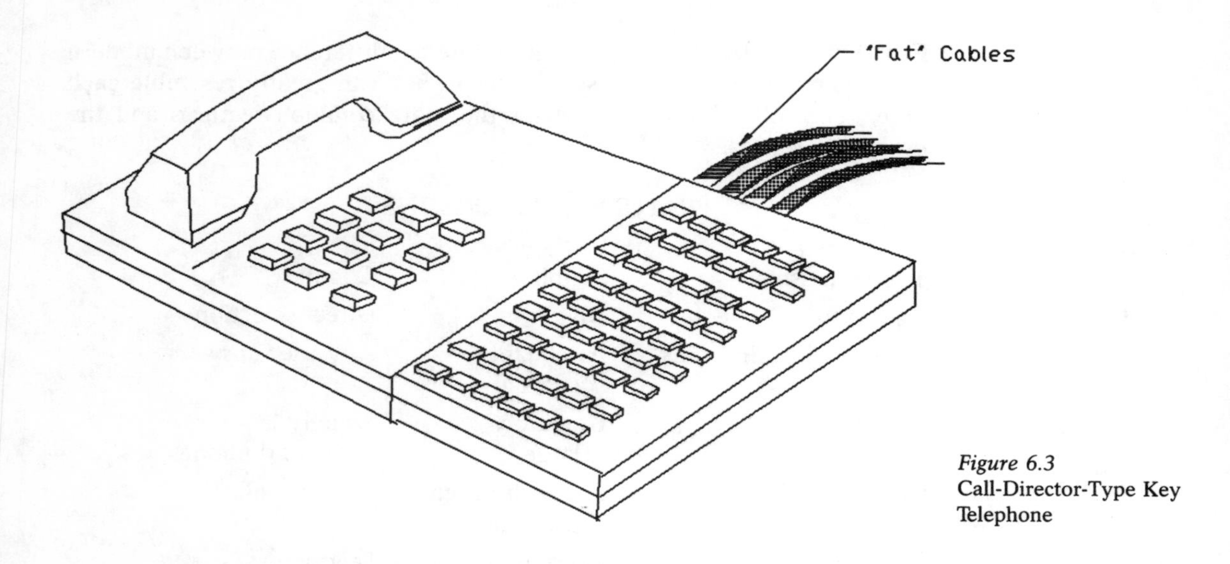

Figure 6.3
Call-Director-Type Key Telephone

sulted in the so-called Call Director® being developed (various other proprietary names were also used). These sets had the capability of accessing from 12-60 or more lines. A large number of bulky cables are required to connect one of these sets.

A modern electronic "key" set is shown in Figure 6.4. Obviously, the electronic key set has a number of features that are not available on the 1A2 6-button set. We will examine some of these features shortly.

In both 1A2 and electronic key systems the station sets are invariably more expensive than the usual sets used on PBXs. (Even so, it is not uncommon to see key equipment connected as PBX extensions).

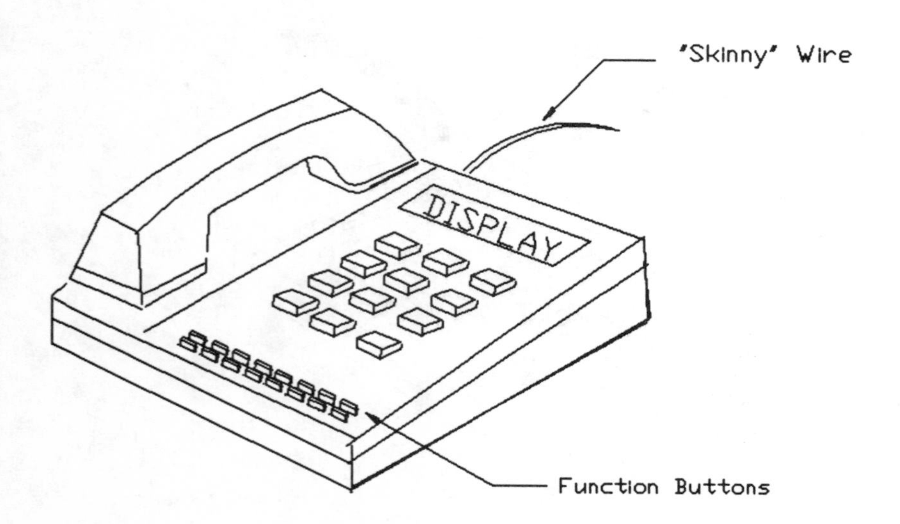

Figure 6.4
Electronic Key Set

Common Equipment PBXs are characterized by having complex central equipment shared by all telephones. The 1A2 key systems have rather simple centralized equipment with *mechanically* complex telephone sets that require that the user provide much of the "intelligence" that a PBX provides.

Electronic key systems generally put a lot of the intelligent logic and control in the telephone set itself; the central control is also electronically complex. The use of this *distributed logic* permits the use of two or three pairs of wires or a "skinny" cable in place of the bulky cables to each telephone, yet allowing full access to many lines from each set. Special data signals are passed between the central control and the telephone sets to establish the connection the user desires.

Features All key systems have one thing in common; the ability to select two or more lines and to be able to put a call on "hold" while using another. Since 1A2 key systems are being phased out, we will not devote more discussion to them. It should be noted, however, that there are hundreds of thousands of these systems still in place.

Electronic Key Systems evolved from simple replacements for 1A2 to a maze of features. In the author's opinion, the most useful and practical features are:

Speed calling, or repertory dialing from a user-inputed database.

Conferencing of inside and outside calls under user control.

Last number redial. A valuable feature when encountering busy numbers.

Queuing for lines that are in contention such as WATS. The system puts the user in a queue and calls user when desired line becomes available.

Message waiting indication on phone that signals that someone left a message.

Paging. This feature permits caller's voice to be heard over a small speaker in phone base to indicate that the called party is requested to speak to the caller. Group paging can be arranged so that all phones will carry the paged announcement.

Toll restriction prevents certain phones from being used for costly toll calls. Various levels of restriction are available.

7 TRANSMISSION II

SIGNALING

Background An early form of signaling that remains to this day is ringing, or sending an alternating current down the line to alert the called end. The familiar telephone bell (or its electronic tone ringer equivalent) is likely to remain for years to come. It's hard to conceive of a simpler arrangement to alert a distant person.

Ring Down Trunks When trunks between central offices were established early in the history of telephony, the same type of signaling was used to alert a distant operator. Instead of a bell, a little latched door-like device called a "drop" was triggered on the switchboard. Figure 7.1 pictures an old switchboard with drops. (To this day the far end of a circuit is called a drop.)

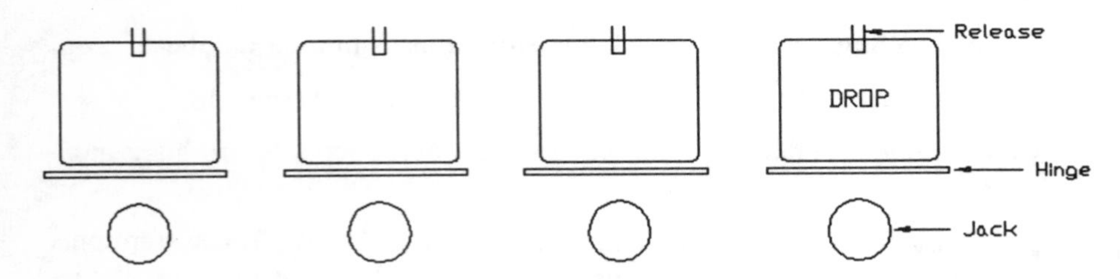

Figure 7.1
Magneto Switchboard "Drops"

Single Frequency Signaling As circuits became longer, it became impractical to continue using standard ringing current to signal, so higher-pitched tones were used to alert the distant operator. So-called *single frequency* or *SF* signaling was born. These tones are detected electronically.

In earlier days, the tone lighted a lamp on the board to alert an operator. SF is in use on many circuits not only to alert equipment but to carry dial pulses. In actuality, tone is placed on an *idle* circuit and removed to indicate seizure of the circuit. In this way constant monitoring of a circuit is taking place.

Multifrequency Signaling Multifrequency (MF) signaling was an invention born from the need for rapid, accurate dialing between distant cities. Five tones are used in combinations of two to indicate ten digits, as follows:

	FREQUENCY				
Digit	700	900	1100	1300	1500
0				X	X
1	X	X			
2	X		X		
3		X	X		
4	X			X	
5		X		X	
6			X	X	
7	X				X
8		X			X
9			X		X

Multifrequency Signaling

Notice that these tones are not the same ones used in Touch-Tone® or DTMF telephones as seen in the table in Chapter 1. DTMF uses two tones out of eight instead of two out of five.

The use of MF signaling predates the use of DTMF by many years since bulky vacuum tubes could be used in the central office environment, where this would be extremely unwieldy at a telephone instrument. Solid-state electronics are used in telephone sets.

E&M Signaling There is probably no subject in telephony that arouses more confusion than the subject of E&M signaling. Unless one works with it every day, there is a likelihood that he or she will forget just what is what.

First of all, there is probably no true meaning for the letters "E" and "M". In the early days of telephony, circuit drawings showed wires being designated with letters, and the designer of E&M apparently had arbitrarily chosen E and M for these particular leads. Perhaps so. Another story has it that E stands for "earth" and M stands for "metallic". In any case, it is useful to think of E as standing for "ear" and M standing for "mouth", as we shall see.

Figure 7.2 (next page) shows an E&M circuit arrangement. When Office A wishes to set up a call to Office B, a voltage condition that is applied to the M lead in Office A arrives in Office B on the E lead; (Mouth to Ear). Office B responds to the summons when it is ready by putting a similar voltage condition on its M lead, which Office A receives on the E lead.*

*Instead of a "voltage condition", SF tones are used on long-haul circuits since ordinary DC voltage cannot be transmitted for long distances.

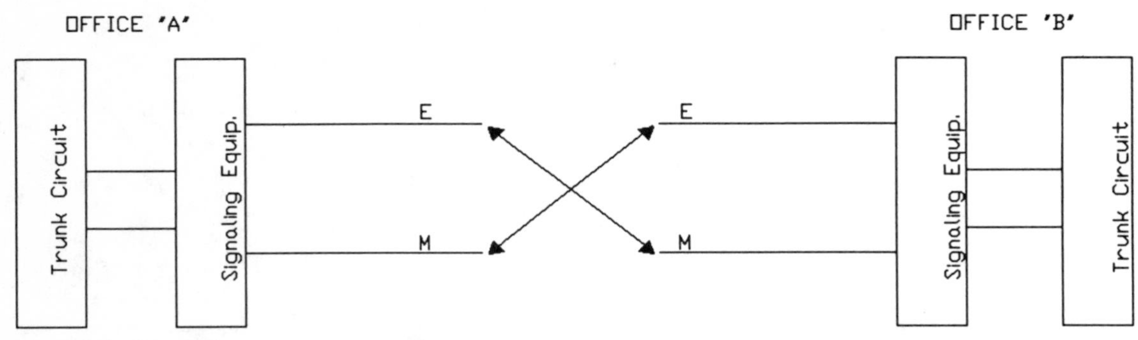

Figure 7.2 E and M Trunk Signaling

Common Channel Interoffice Signaling - CCIS As the number of central offices and toll offices utilizing time-division switching has increased, there has been an increase in the use of CCIS. CCIS and time-division switching and transmission go hand in hand.

Figure 7.3 is a diagram of a CCIS system. CCIS is a system for exchanging information between electronic toll and central offices through specialized signaling trunks. The information is sent as high-speed digital data. All information related to the setup and takedown of trunks and addressing (dialed number) is exchanged by means of the signaling trunks. Signaling for many trunks is carried by one signaling trunk. In this way, the voice circuits are not used to transmit this information.

Calls are set up and taken down much faster. This can result in fewer trunks being required in large trunk groups. The users are pleased because calls are completed faster after dialing. In addition, it is not even necessary for a voice path to be set up if the called number is busy or does not answer. Local originating offices can return the busy signal or ringing

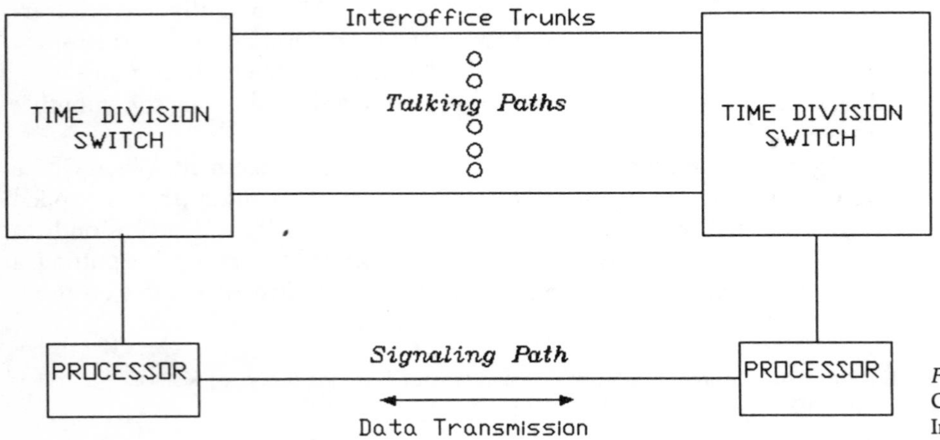

Figure 7.3 Common Channel Interoffice Signaling

tone. The latter feature is most valuable on more expensive circuits, such as international.

Many other features are potentially available with CCIS. To name a few:

Long distance call tracing is made relatively easy.

Display of calling party's telephone number at called party's phone before answering.

"Follow Me" service to temporarily forward calls to a different number.

CARRIER SYSTEMS

We have already discussed one of the more advanced systems to carry more than one conversation over wires using the PCM technique, but there is still a significant amount of non-digitized facilities in intercity telephone plant. As may be suspected, non-digitized, or analog systems existed long before digital systems since analog is the "natural" form of voice transmission.

Modulation Simply stated, modulation is a means of impressing one signal on another in such a way that one signal is "carrying" another. A fine example is the human voice. When air is expelled from the lungs it can be made to vibrate the vocal cords and produce a tone. Unless we start talking in Morse Code, little information can be conveyed by this tone; however, we cleverly *modulate* the voice tone with our mouth and tongue to produce speech. We are varying the *amplitude* of the voice tone. The signal that is being varied is called a *carrier.* In the case of human speech, the tone produced by the vocal cords is the *carrier*.

There are three common ways that one signal can be caused to modulate another:

The amplitude (or instantaneous loudness) of the carrier is varied with the signal being carried. Figure 7.4 (next page) illustrates this method of modulation.

The frequency is varied.

The instantaneous phase is varied.

Multiplexing As shown on the diagram in Figure 7.5 (page 45), simultaneously impressing more than one signal on a carrier to carry many conversations on a single facility is called *multiplexing*. All three modulation techniques have been used in commercial telephone analog carrier systems. Hundreds of simultaneous conversations can be multiplexed on a single facility such as microwave or coaxial cable systems. Recall the

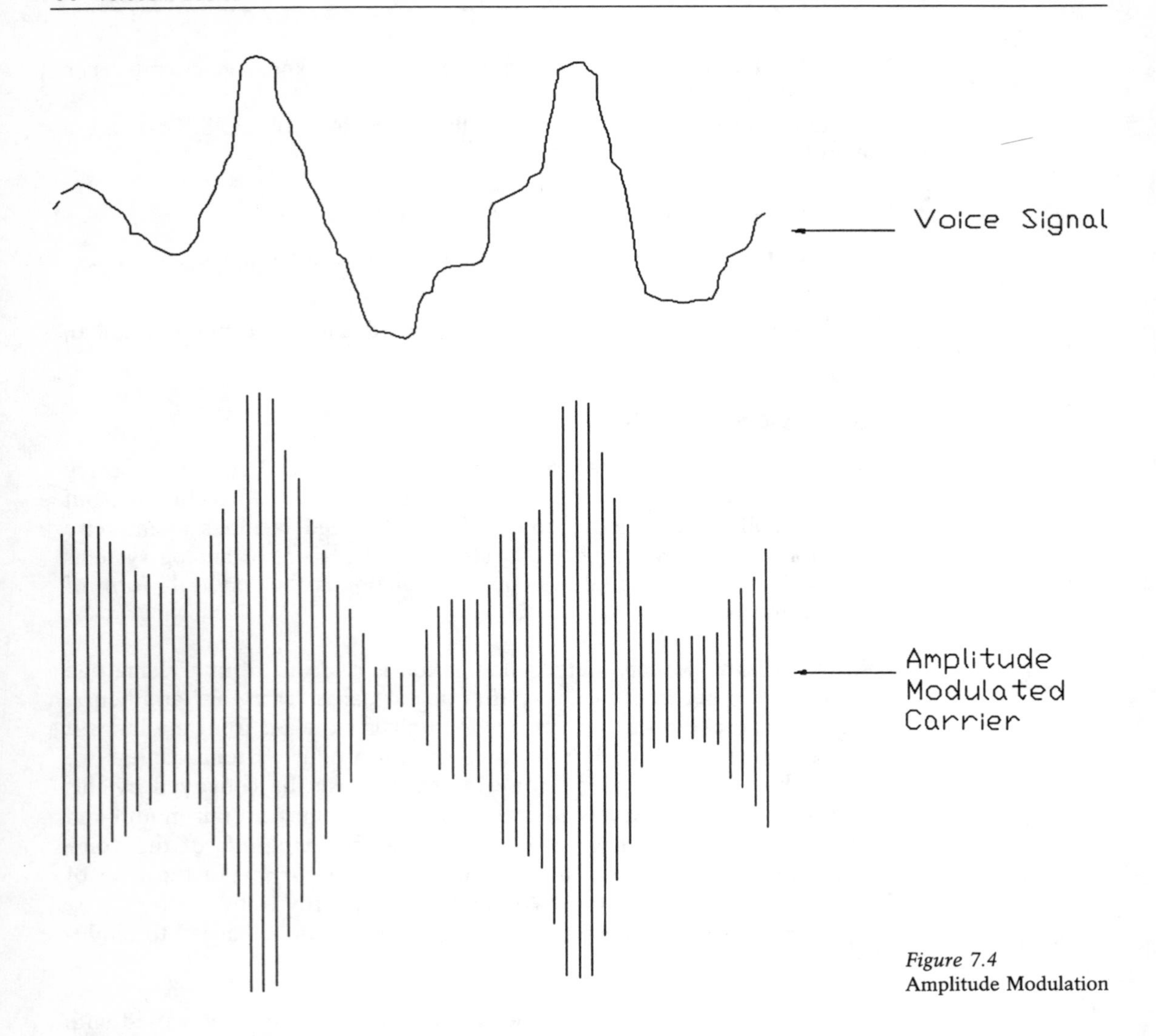

Figure 7.4
Amplitude Modulation

time division multiplexing system called T1 that we briefly discussed in Chapter 4. In this system, 24 voice channels are multiplexed on ordinary telephone wires using binary pulses.

Transmission Facilities A variety of facilities are used to transmit the carrier signals. We have mentioned three briefly. Let us now examine all of them in a little more detail.

Common *telephone wire* found in cables is always in the form of pairs of wire twisted together. The twisting is necessary to keep the separate wires from acting like an antenna. Interference induced in one of the wires of a pair also is induced in the second wire in an opposite manner.

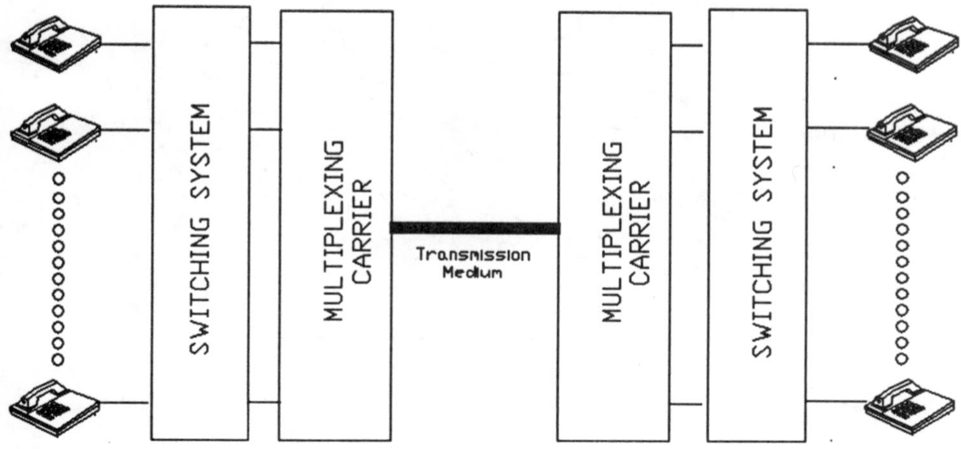

Figure 7.5 Multiplexing Carrier System

Coaxial cable was developed to provide a more effective way to isolate wires from outside influences such as electrical noise or radio signals. As the name implies, the conductors share a common axis. Figure 7.6 (next page) illustrates this type of cable. The reader has probably become familiar with this type of cable with the advent of VCRs and other modern devices that use "coax" for interconnection of television signals. Coax has a further advantage; electrical characteristics, such as impedance can be accurately controlled.

Microwave radio systems are another immensely popular means of carrying telephone signals. The name "microwave" was derived from the fact that the *wavelength* of these radio waves is very short. Because the wavelength is short, these radio waves take on many of the characteristics of light, which has exceedingly short waves. Unlike normal radio broadcasting, microwave transmission is very directional. See Figure 7.7 (next page). One antenna (or "dish") can be directly aimed at another. In this way, there is very little potential for interference from another signal and very low power is required to send signals from point to point. Microwave transmission has the potential to carry a thousand or more multiplexed voice signals.

Satellite transmission is a specialized form of microwave transmission where the "tower is in the sky". Refer to Figure 7.8 (next page). All communications satellites are in a *geosynchronous* orbit. This means that the altitude and speed are such that the satellite is circling the globe at the same speed that the earth rotates, so it appears to be stationary. This happens to occur at 22,000 miles above the earth at the equator. Even though radio waves travel very fast (186,000 miles/second), the round trip to a satellite takes about 1/4 second. This is enough to be

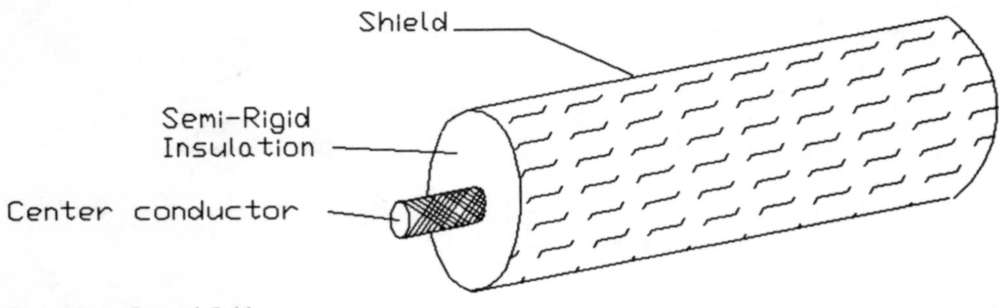

Figure 7.6 Coaxial Cable

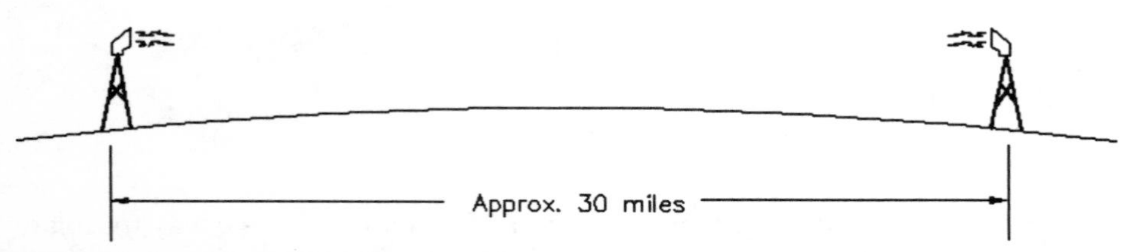

Figure 7.7 Microwave Transmission System

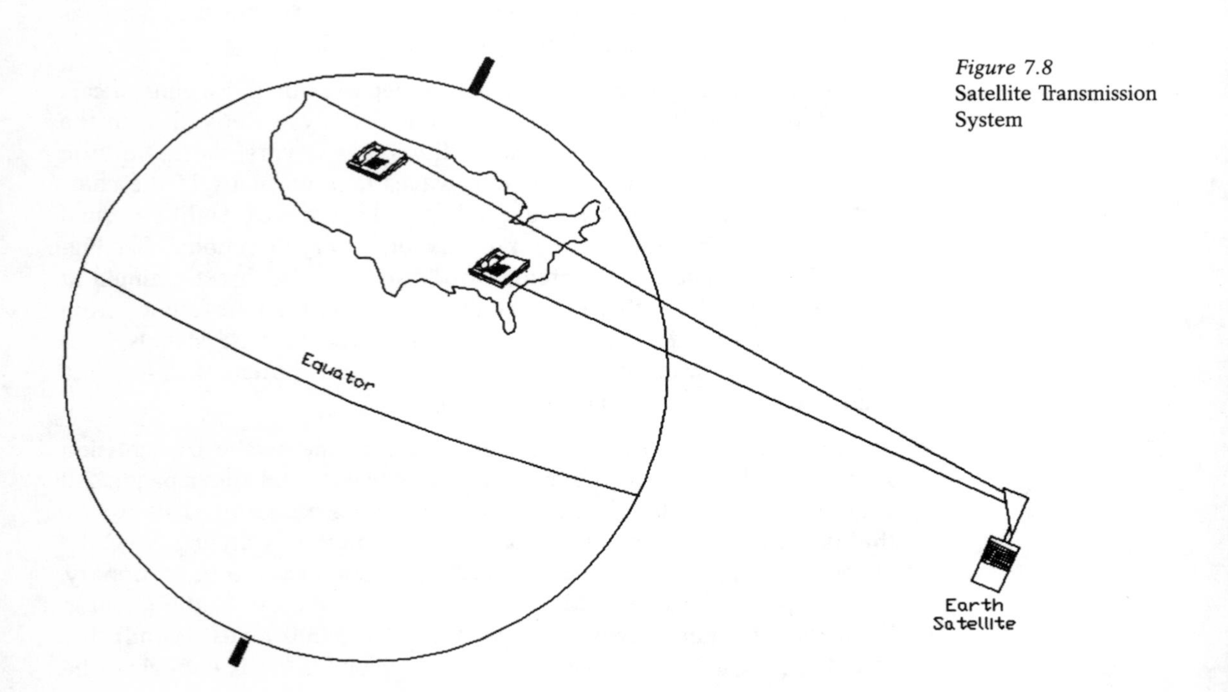

Figure 7.8
Satellite Transmission System

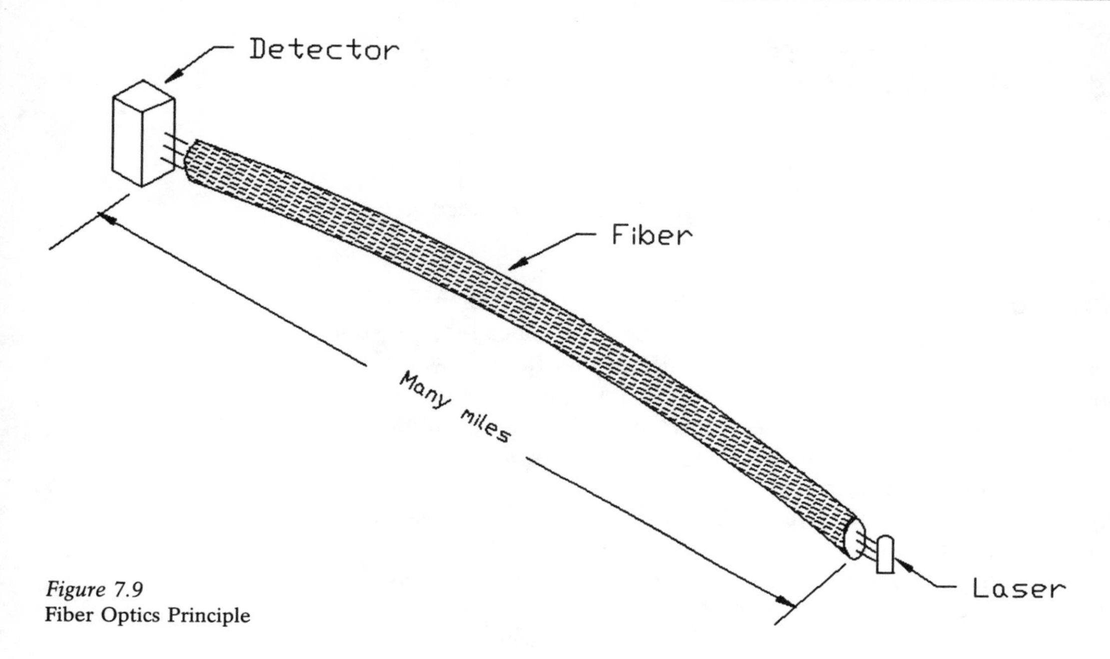

Figure 7.9
Fiber Optics Principle

noticeable at times to callers and can be disasterous to data transmission. For this reason, most long distance connections are arranged to avoid multiple satellite hops. However, they can be encountered on some international calls.

Fiber optics are the latest form of facilities to be used for transmission systems. Refer to Figure 7.9. Light, not radio waves, is transmitted through glass fibers about the size of a human hair. Several fibers are bundled in a cable. Light is turned on and off *billions* of times per second to represent the binary signals. (Recall that in the T1 system that the binary rate was 1.5 *million* bits per second). In this way, thousands of simultaneous voice signals can be transmitted over a single fiber. Analog carrier systems cannot be used on fiber.

8 FACILITIES

This chapter deals with a variety of topics that have been generally classified as "facilities".

DDD Hierarchy Back in the days before Direct Distance Dialing (DDD) by callers was the norm, operators manually placed long distance calls for customers. A special book called, *Rates and Routes,* was used by the operator to determine how a call was to be routed through the network. It wasn't unusual for five or more intermediate operators to be involved in a call to out-of-the-way, small towns. (In those days it was fascinating to listen-in as a call was placed.)

With the advent of DDD a definite plan had to be developed in conjuction with "smart" switches so that calls could be automatically routed. The so-called "DDD Hierarchy" was what evolved. Refer to Figure 8.1.

A layered hierarchy was developed in which "lower class" switches home on their higher class associates which are geographically distributed. Generally, the higher class switches are located in larger cities.

Each of the classes is assigned a number (and designation) as shown in the diagram. In the usual case, local telephone customers home on a Class

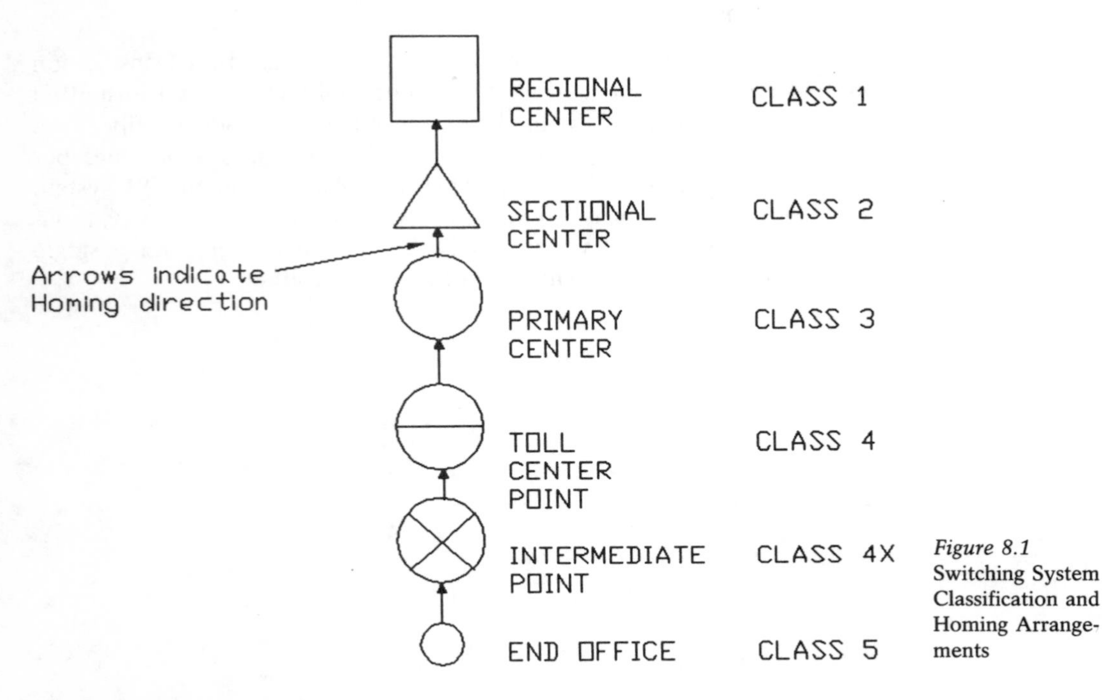

Figure 8.1 Switching System Classification and Homing Arrangements

5, or End Office. In turn, each class homes onto and is connected to the next higher class by circuits called "Final Trunk Group" (FTG) trunks.

There are always FTGs provided between Class 1 (Regional Center) switches, since these are the absolute last choice a call has to be completed. FTGs are designed to have a very low probability of call blocking; that is, enough circuits are provided to assure an excellent chance of call completion.

Refer now to Figure 8.2. We will examine how a call from a caller at (716) 346-3057 to telephone (813) 525-2098 could be completed.

> The 716 caller dials into his serving (Class 5) office.
>
> The call is passed to the Class 3 office via the Classes 5-4X-4 offices since this is the only possible routing provided.
>
> At the Class 3 switch there is circuit (dashed) that is a High Usage Trunk (HUT) to the desired path to the 813 customers. *This is a first choice route* for this call. However, if all of the HUTs are busy, the call would be passed on to the Class 2 switch.
>
> At the Class 2 point, there is again a HUT choice available to the path to 813 customers via the Class 2 office in this path. If these are all busy,

Figure 8.2 DDD Alternative Routing Example

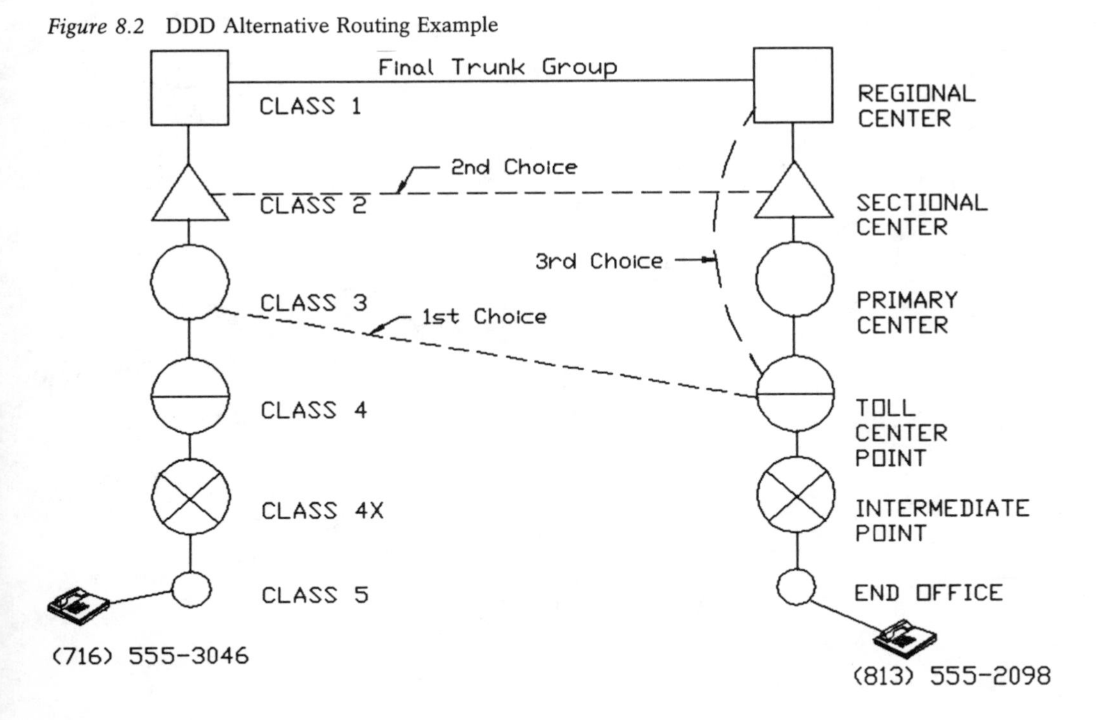

the call will be passed to the Class 1, where an FTG will be used to send the call to the appropriate Class 1 (Regional) office.

At this Class 1 office, a wide-spaced dashed HUT is shown which would be first choice for completion via the Class 4 (Toll Center) servicing the called party's central office. Again, if the HUTs are busy, the call would proceed down via Classes 2-3-4-4X to 5 for completion.

All of these actions will take place in seconds without the caller being aware of the complex choice examined for his call's completion.

Figure 8.3 is a diagram that shows typical switching system groupings by office class.

Rates and Tariffs Strictly speaking, rates and tariffs are not facilities, but they are related to them so intimately that this is an appropriate time to discuss them.

A tariff is nothing more than a schedule of charges and stated obligations of the provider of service.

In the days before the Carterfone decision, things were quite a bit less complex. For the most part, the Bell System provided all the services a user might want, from telephone sets to long distance service, and it was Bell's *exclusive* province. If a particular service was needed, one simply

Figure 8.3 Typical DDD Switch Groupings

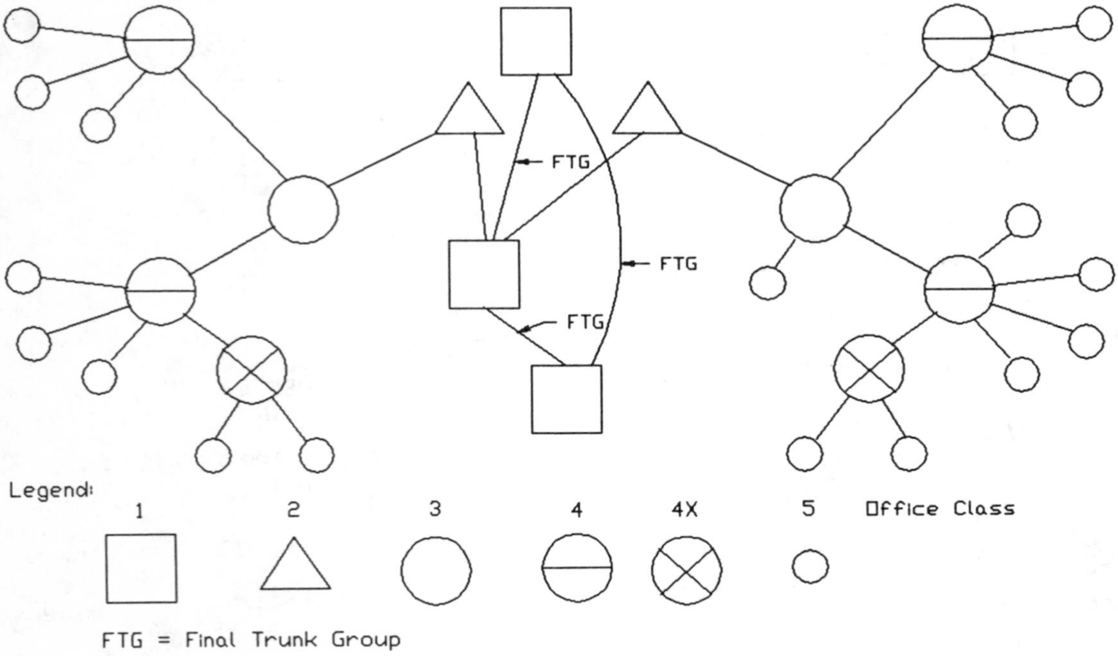

called the “phone company” who took care of everything—albeit slowly and reluctantly in some instances. Virtually everything that was provided was tariffed. In some cases, “special assemblies” were arranged and provided at a fixed rate.

A tariff is *not* simply based on the cost of a service, but its perceived value as well. In this way some items provide a very good rate of return, while others are, in effect, subsidized. (Because of this, after divestiture of the Bell System, the remaining local operating telephone companies have had to increase the rates for local service which had been subsidized by long distance service.)

Tariffs are submitted for approval to State or Federal Communications Commissions for approval before they can become effective. Not infrequently, there is a period of negotiation between the telecommunication companies and the authorities.

Overseas Facilities Making a call overseas was once very expensive. Back in the forties the cost to call any overseas country was $12.00 for three minutes. Shortwave radio was used to provide the transmission, and it was noisy and undependable as well as providing very few channels.

There were a number of undersea *telegraph* cables in place as far back as the early 1900s, but these could not carry voice because of technical considerations. An historical item: The first undersea telegraph cable worked only a few hours after it was placed because of seepage of sea water into it.

The advent of the first undersea *telephone* cable in 1955 was a major milestone in international calling. Even though it initially carried only 24 simultaneous calls, it was an enormous advance in the technology. Vacuum tube repeaters with a life expectancy of 20 years were laid along with the cable every 20 miles or so. The power for the repeaters was supplied from the terminus points in Maine and Falmouth, England.

An advance was rapidly made to derive many additional cable circuits and is called TASI, or Time-Assignment-Speech-Interpolation. See Figure 8.4. TASI takes advantage of the fact that talkers generally talk in spurts and only one at a time. The idle time is used to send spurts of other conversations along a circuit. In this way, conversations are temporarily assigned an available circuit when needed. TASI is only viable when the facility cost is very high, such as in undersea cables.

Since 1955 there have been many additional undersea cables placed from the US/Canada to Europe and Asia. Of course, as solid-state electronics became more dependable they were used instead of vacuum tubes for the repeaters. All of the cables to date use coaxial copper facilities. There will shortly be a cable in place that uses fiber optic (glass) facilities. This, of course, will greatly increase the number of circuits per cable as well as providing very high-speed data transmission capability.

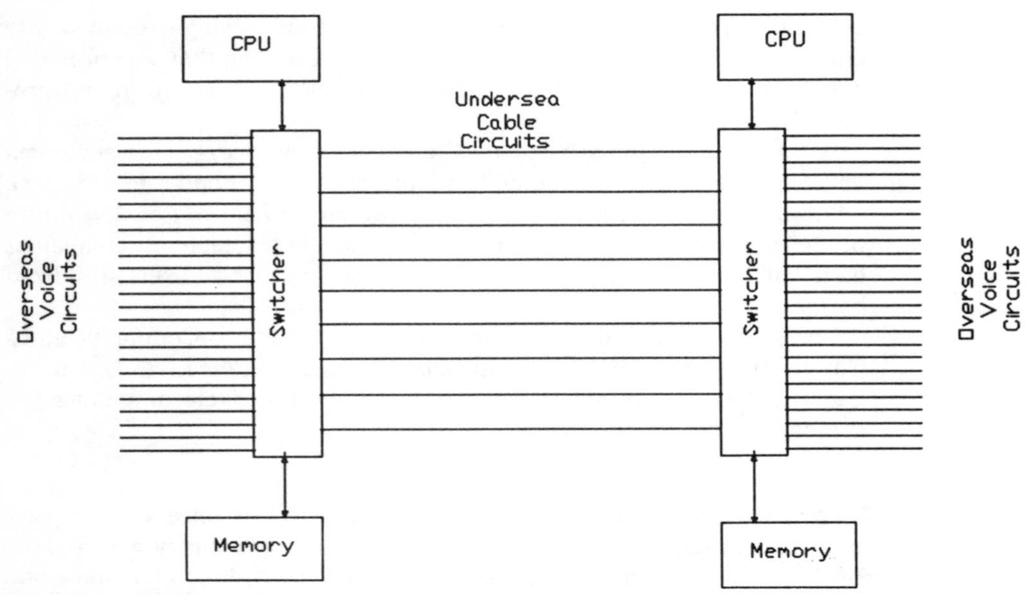

Figure 8.4 Time-Assignment-Speech-Interpolation

In addition to undersea cable there are a number of satellites providing overseas transmission in competition with cable. As further fiber cables are placed, it is likely that the principal function of the satellites will be to provide circuits to less populated countries.

Cellular Phones Even though police cars didn't have two-way radios until the 1930s, shortly thereafter there was demand for telephone service in private cars. Of course, the only possible way to provide such service is by means of radio. As long as only a relatively few needed mobile telephones, it wasn't too difficult to provide service in larger communities. However, it wasn't long before the demand exceeded what could be provided without tremendous contention for service. In most cases, just one or two channels could be made available to serve hundreds of users. Furthermore, all of the service had to be handled by operators—it just wasn't practical to provide dial service, especially with the great contention for service. The obvious solution was to provide more channels, but there is just so much radio frequency space, and the mobile telephone users were only one of the contenders for the space. As a result, the service was downright awful.

Provision of improved mobile telephone service was a problem that cried out for a solution. Two developments made it possible:

Advanced techniques in UHF radio.

Computers.

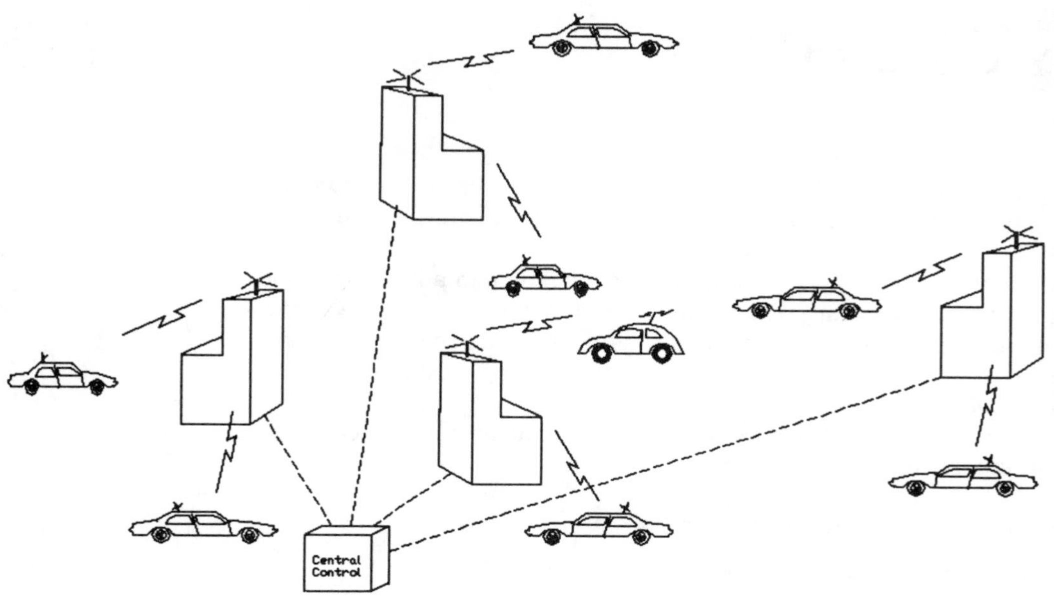

Figure 8.5 Cellular Telephone System

Space research accelerated the development of solid state electronics capable of working with Ultra High Frequency (UHF) radio signals at a reasonable price. UHF inherently provides much larger bandwidth, and larger bandwidth means more channels for mobile service.

More channels *per se* would only temporarily alleviate the problem of crowded channels. What was needed was a way to allocate channels "on the fly" to users as they spoke. Figure 8.5 illustrates the solution—*cellular radio.*

Very intelligent mobile telephones are utilized and transmitter/receivers of the radio signals are located in "cells" that divide the serving area. Since UHF signals are very directional, properly located receivers can serve only small sectors of a large city, and as cars move about, the car's signal is "handed off" to the receiver that is getting the strongest signal. This can only be done with a network system centrally controlled by a computer. All of this means that many more users can be simultaneously served in a locale.

Most car units (and hand-helds) have advanced features such as memory dialers, speaker-phones, etc.

9 TRAFFIC

Going to the Bank The experience of going to a bank is a down-to-earth example of traffic design and it is quite analogous to the telephone situation, so we will start with it.

Until recently, it was almost inevitable that standing in a line at a bank was going to be frustrating. You would likely get into a line that was the slowest moving. Fortunately, most banks have an arrangement nowadays where there is a single line that is arranged to assign customers to the next available teller. See Figure 9.1.

There are a number of factors that the bank manager has to consider when laying out the service:

- Number of teller positions
- Number of customers expected in an *average busy hour*
- How long customers will tolerate waiting

Figure 9.1 People in Queue at a Bank

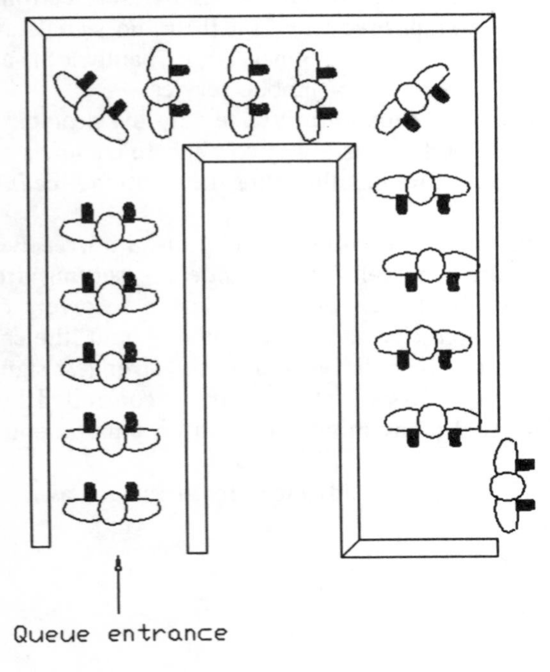

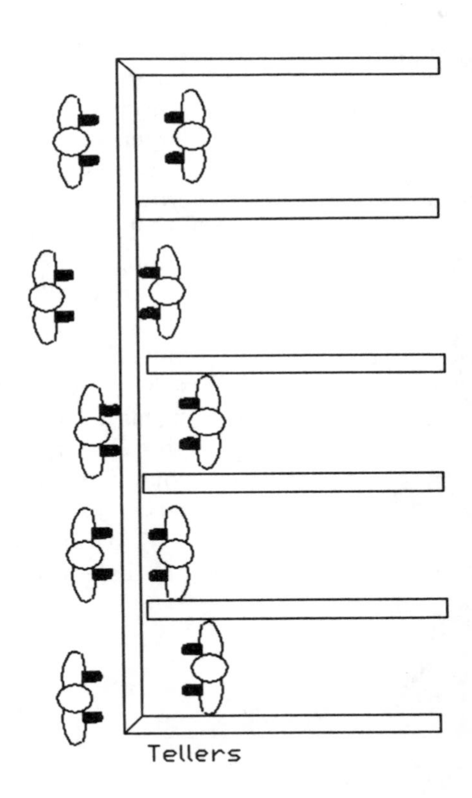

- Average length of time per transaction
- Time of day when they are the busiest—the busy hour

Naturally enough, economic decisions must be made when considering these factors. From the customer's perception, ideally, he should never have to wait at all. The manager must make some compromises to accommodate the bank's traffic in a reasonable way. The compromises inevitably will cause some *delay* to the customers during the busy hour(s). His primary problem is to determine how many tellers are required.

Busy Hour The busy hour (BH) is very important in the analysis of traffic. First, understand what it *is not.* It does not mean the busiest hour that might possibly be encountered at any time in the year. Busy hour is almost always taken to mean the *average* busiest hour on an average day. In the case of the bank, the manager cannot design the facilities to handle unusual, high traffic requirements; rather, he must plan to accommodate the usual requirements. Also, it is not unusual for there to be more than one busy hour in a day. Shortly, we will see how the busy hour traffic will be used to determine how many tellers should be on duty during the busy hour.

Holding Time Another important factor is the average time it takes to serve a customer—the *holding time.*

Acceptable Delay The next factor that our manager must determine is the length of time, or delay, that his customers will tolerate standing in line. Again, the ideal is no (zero) time. However, almost everyone is willing to wait a reasonable length of time to be served.

Solving the Problem Once all of these facts are in hand there are standard formulas or tables that can be used to see how many tellers are needed during the busy hour. Notice again that all of the factors are based on *averages*. It happens that this is just fine, since the calculations will be statistical and all statistics are based on averages or means.

For our purposes we are going to assume some values for the manager to see how many tellers he must assign during the BH. Customer Arrival Rates illustrated in Figure 9.2 were arbitrarily compiled for each half-hour. The busy hour can be seen to be from 10 to 11 AM.

Average holding time	5 minutes
Acceptable delay	10 minutes
Arrival rate	36 customers per minute

The diagram in Figure 9.3 will be used to see how many tellers the manager must assign in order to meet these criteria. The scale of the

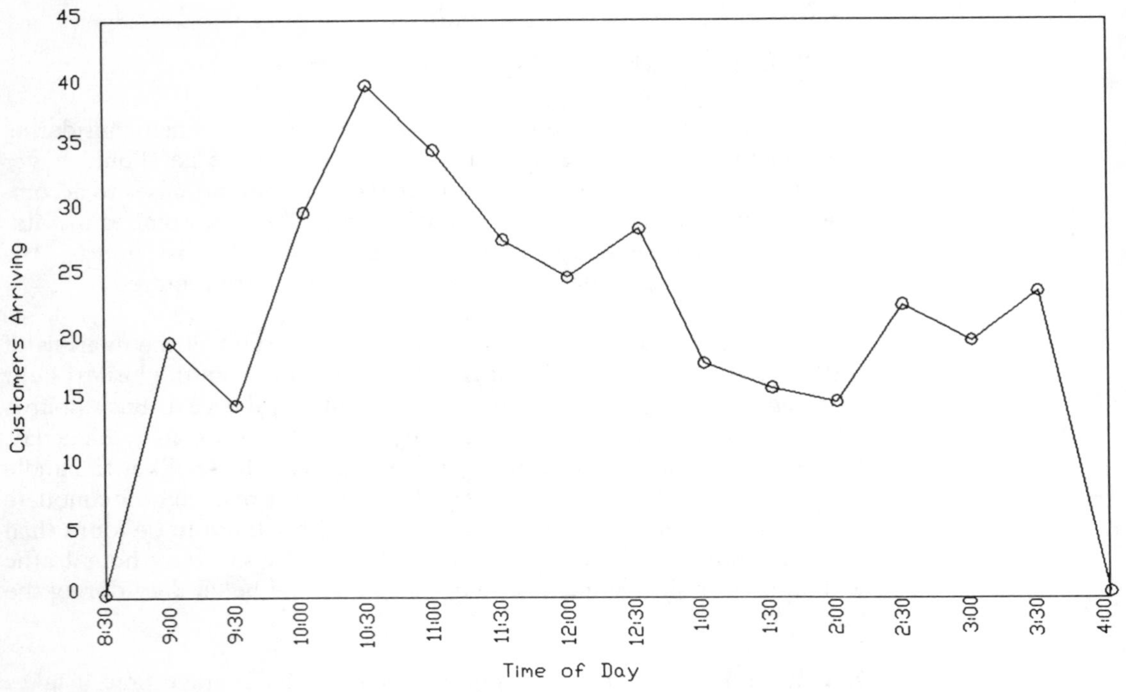

Figure 9.2 Histogram of Bank Customer Arrivals

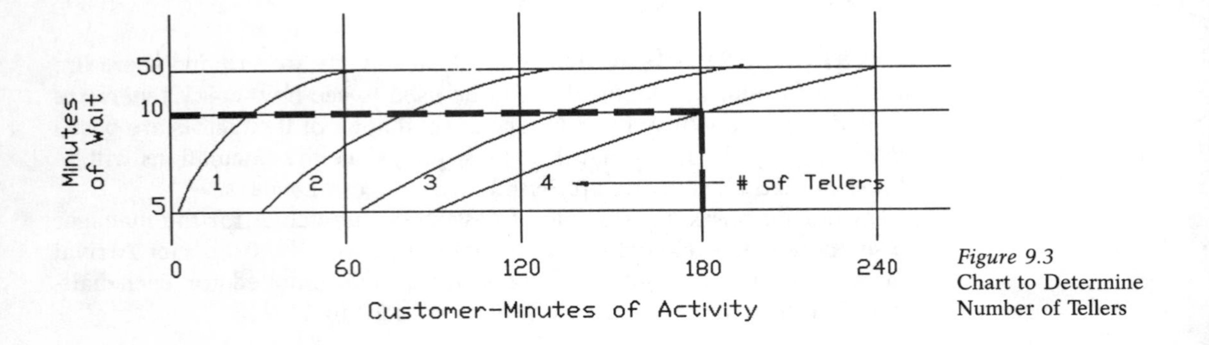

Figure 9.3
Chart to Determine Number of Tellers

"Minutes of Wait" is logarithmic, so the numbers increase rapidly with the distance along the scale.

Solution The amount of work the tellers must perform is obtained by multiplying the number of customers arriving by the average time it takes to serve each of them.

36 customers/min × 5 minutes/cust = 180 customer-minutes

The acceptable delay is given as 10 minutes.

By following the line for 10 minutes' delay to its intersection with 180 customer-minutes, we see that 4 tellers are needed to do the job.

There are some other interesting things to see while we are examining this diagram:

If the customers would accept a 40-minute delay, there would only be a need for 3 tellers. A surprising result! Adding just one teller reduces the delay by a factor of 4. In fact, up to a point, adding extra facilities (such as a teller) increases the efficiency of the system in an exponential manner. Changing the number of tellers dramatically affects the waiting time for the customers.

Reducing the arrival rate of customers to about 90 per hour would result in an average waiting time of 5 minutes.*

The bank example was used to introduce the concept of *traffic*. We will now continue with traffic concepts as they apply to telephony.

Let us now turn our attention to telephony examples.

Queuing As was discussed in an earlier chapter, in the early days of telephony people trying to make a call were at the mercy of telephone operators. Some interval after lifting the receiver an operator would respond with, "number please?" Of course, the interval would be dependent on the number of operators on duty. This is analogous to the bank teller situation in many ways. Users were in a queue for service.

We are still subject to queuing today, but it is usually not so obvious. Getting dial tone from a modern central office or PBX also puts us into a queuing situation. There must be a limited number of facilities that provide dial tone and accept the dialed digits, and these facilities must be traffic-engineered in the same manner using the same type of diagrams we used above.

Another queuing occasion familiar to many business people is attempting to call out on a WATS (wide area telephone service) line. Because of their cost, most businesses have only a limited number of these lines;

*The diagram reproduced in Figure 9.3 is only a portion of a much larger one called an "Erlang C" diagram. It was specially adapted here for the bank example from diagrams designed for telephony use. However, the adaptation is quite valid as long as the same conditions such as "random" arrival of customers apply.

accordingly, it is common for them to be accessed by means of PBX *trunk queuing.* Factors such as the average holding time, average calling rate and acceptable delay are used to engineer these trunks.

CCS There is another term that must be introduced that is unique to telecommunications. Unfortunately, it is somewhat cumbersome, but it is so widely used that it must be defined: CCS is used to describe user activity in telephony. (In this way it is akin to the term, "customer-minutes" used in the bank example.) CCS stands for one-hundred-call-seconds—the initial C is the Roman numeral for 100. Readers that are not from North America will note that 36 CCS is equivalent to an *erlang.*

CCS is a mathematical expression that can describe a number of combinations of situations of time usage. For example, 10 CCS can be the result of any of the following:

10 calls of 100 seconds each
100 calls of 10 seconds each
1000 calls of 1 second each
1 call of 1000 seconds

Although this may be somewhat confusing, it is a very useful way to describe traffic in a PBX or central office. Both the Poisson and Erlang B Tables we will be examining require the use of the CCS designation to determine the trunking.

Poisson Tables A new situation is to be encountered here. Instead of callers being patient and remaining in a queue for service, they are expected to temporarily give up and try some time later. The most common example of this is when dialing 9 for an outside line from a PBX. Just so many 9th level lines can be economically provided, so somebody occasionally has to be thwarted in the attempt to make a call during the busy hour(s).

Speaking of busy hours, examine Figure 9.4. This diagram is a "histogram" of typical calling activity in a large business environment. As expected, there is very little activity prior to 8 AM. Calling picks up rapidly until a peak occurs at 10 AM. In fact from about 9:30 until 10:30 there is almost a sustained peak. A smaller peak in calling activity occurs from 2 to 3 PM. Of course, this is only a representative curve; businesses do vary. The busiest day of the week is usually Monday. In order to be certain, it is highly advisable to sample actual calling activity; however, a rule-of-thumb is to assume 17% of a day's calling occurs in the busy hour, whenever it may be.

One more term must be introduced before proceeding with the Poisson calculations: Probability of Blocking or "P". This simply is a term to indicate acceptable service. A value of P.02 means that 2 calls in 100 will

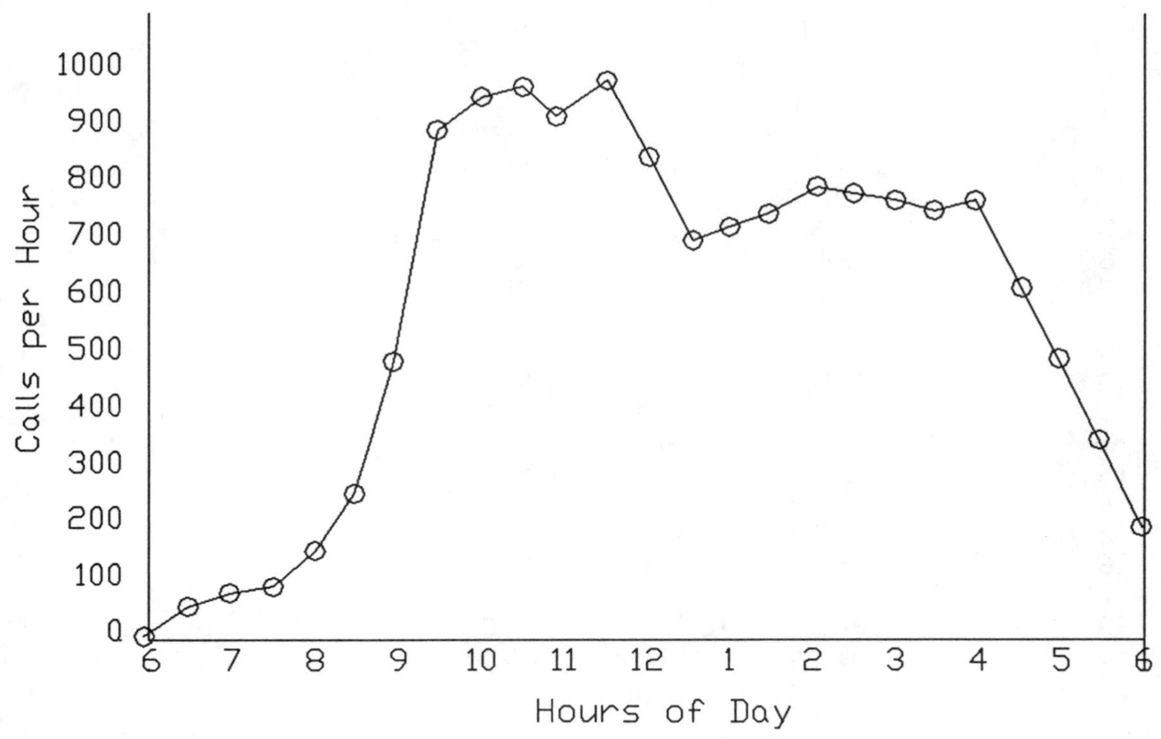

Figure 9.4 Histogram of Typical Call Arrivals

be blocked in attempt to obtain service *and they will hang up and try later* in the case of using the Poisson tables.

All traffic calculations are made using the busy hour data.

To illustrate the use of Poisson Tables to determine the number of trunks needed for a PBX application, let's use the following parameters:

100 phones in the system

average of 5 minutes of "outside" calling per phone in the busy hour

desired probability of blocking is P.02

CCS is first calculated by the following:

$$\frac{100 \text{ phones} \times 5 \text{ minutes} \times 60 \text{ seconds}}{100} = 300 \text{ CCS}$$

This is all the information needed to determine the number of trunks that are needed. Look at a portion of a Poisson Table reproduced in Figure 9.5 (next page):

Trunks	P.01	P.02	P.03
1	0.4	0.7	1.1
2	5.4	7.9	9.7
3	15.7	20.4	24.0
4	29.6	36.7	41.6
5	46.1	55.8	66.6
6	64.4	76.0	82.8
7	83.9	96.8	105.0
8	105.0	119.0	129.0
9	126.0	142.0	153.0
10	149.0	166.0	178.0
11	172.0	191.0	204.0
12	195.0	216.0	230.0
13	220.0	241.0	256.0
14	244.0	267.0	283.0
15	269.0	293.0	310.0
16	294.0	320.0	337.0
17	320.0	347.0	365.0
18	346.0	374.0	392.0
19	373.0	401.0	420.0

Figure 9.5
Poisson Traffic Table

Follow down the P.02 column until a number equal to or greater than 300 CCS is found. In this case the number 320 exceeds 300, but the 293 one row higher was too small. Looking to the far-left column in the 320 row, we find 16. Hence, 16 trunks can accommodate this offered traffic. In actuality, the service would have slightly less blocking than P.02, also.

Let's assume that P.01 service was required by this company. How many trunks would be required then? As we can see this would require 17 trunks. Similarly, if the user was pleased to have 3 of 100 calls blocked in the busy hour, we can see that only 15 trunks would be needed.

While we are looking at this portion of a Poisson Table, we should note some interesting things:

Up to a point, each added trunk can carry more traffic. Note the increase from 1 to 2 and then 2 to 3 in any column. The increased carrying capacity is dramatic.

As trunks are added, each trunk then carries more load. However, this small portion of a Poisson Table doesn't show that as the trunk group becomes very large (i.e., around 30) that the increasing efficiency levels off. In any case, the larger the trunk group can be, the more efficient it becomes up to a certain point. This illustrates the desirability of making groups as large as possible; for example, when selecting WATS bands by automatic route selection (ARS) software. Having a small number of small trunk groups is not a good idea!*

Erlang B One final traffic calculation should be considered: Where, instead of callers hanging up when all circuits are busy, they are automatically sent to another route. A good example of this is when a business has multiple-choice WATS access. If the most economical WATS lines are busy, the call is routed to the next most economic group of WATS lines or 9th level trunks for completion. Traffic analysis for this circumstance should be made with Erlang B (blocked calls not dropped) Tables.

Figure 9.6 is an Erlang B table. It is different from the Poisson Table in a number of ways. Note that probability of blocking is not a part of the table. Instead, the "carried" and "overflow" CCS for each of several trunks is shown. Let's use an example to see the use of the table:

Offered load 200 CCS

Assume 5 WATS lines

The trunks will be accessed in a rotary fashion

Following 200 CCS across the table we see:

that the first line (trunk) can be expected to carry 31 CCS (out of a theoretical 36 CCS maximum) and overflow the remaining 169 CCS to the next line

the second line will carry 29 CCS and overflow 140

the third line will carry 27 CCS and overflow 113

et cetera until the 5th line carries 22 CCS and overflows 65 CCS

Figure 9.6 Erlang B Traffic Table

	Quantities of Trunks																			
	1		2		3		4		5		6		7		8		9		10	
CCS Offered	Car.	Ofl.	Car.	Ofl.	Car.	Ofl.	Car.	Ofl.	Car.	Ofl.	Car.	Ofl.	Car.	Ofl.	Car.	Ofl.	Car.	Ofl.	Car.	Ofl.
198	31	167	29	138	27	111	25	86	22	64	19	45	15	30	11	19	8	11	5	6
200	31	169	29	140	27	113	25	88	22	65	19	47	16	31	12	19	8	11	5	6
202	31	171	29	142	27	115	25	90	23	67	19	48	16	32	12	20	8	12	5	7

*There are a number of rules-of-thumb that have evolved for simple traffic calculations that can be seen to be inaccurate. For example the rule of seven PBX lines per CO trunk is frequently used. In our example, we can see that the resulting 14 trunks would provide less than the desired service. Be careful in using old rules-of-thumb.

It is assumed that the overflow 65 CCS will either go to another WATS group or to CO lines. Note that if no provision is made for overflow from the 5th line, that 65/200 of the calls would be blocked, resulting in P.325 service—very poor. Economic calculations including the cost of lines can be used to determine just how many WATS lines (in this case) should be in use.

Other Methods We have just lightly covered the subject of traffic, keeping in line with the concept of this book being a primer. Because of the importance in determining the quantity of expensive circuits to provide, there has been a lot of effort in the development of more accurate means to forecast traffic requirements. Accordingly, there are a number of other traffic calculation devices available that are more advanced than the Poisson and Erlang Tables.

10 DATA TRANSMISSION I

BACKGROUND

In reality, data transmission on wires is not a recent development. The telegraph system was invented long before telephony, and telegraphy is truly a rudimentary form of data transmission.

As we will see, data transmission is the conveyance of ON and OFF signals—this is *precisely* what telegraphy is. In its simplest form a series of clicks are transmitted that conforms with Morse Code. Look at Figure 10.1. A "key" completes a circuit to a "sounder" at a distant point, which responds to the current flow by producing a click that *roughly* corresponds with the key depression. (Note that the same "smearing" discussed in Chapter 4 causes the pulses that arrive at the sounder to be distorted somewhat.)

Our interest at this time is in the transmission of data signals similar to the telegraphic ones, but at a much greater speed than telegraphic signals, as we will see later.

FUNDAMENTALS

Binary Binary refers to two-state. It is best to use a few examples as to what is meant by "two-state":

ON	**OFF**		
One	**Zero**	**Hi volts**	**Low volts**
Hole	**No-Hole**	**Current**	**No-current**

Figure 10.1 Basic Telegraph System

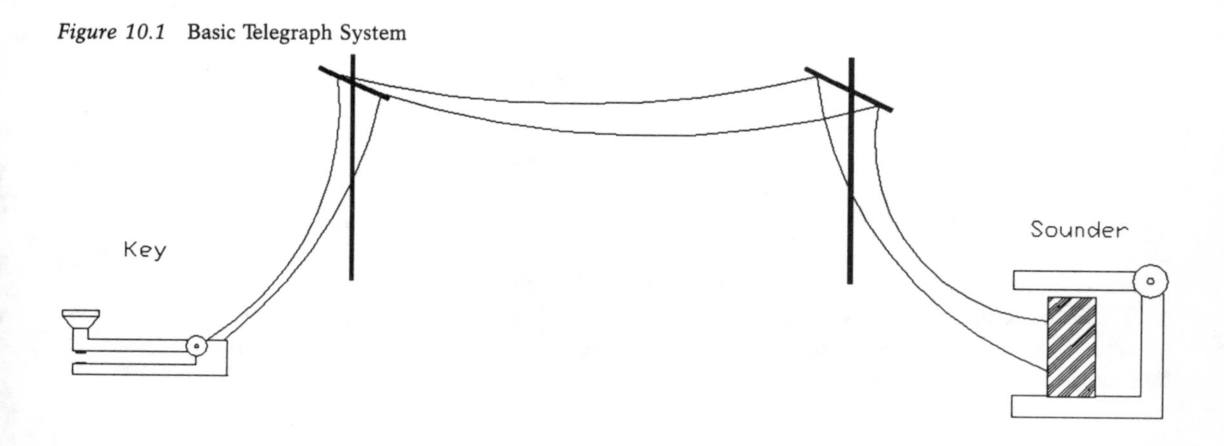

The familiar wall switch is a good example of a binary device. It is either ON or OFF. Other devices that can be in two distinct states are:

Light bulb

Electromechanical relay

Transistor or other solid-state device

Bits "Bit" is a contraction of *Bi*NARY DIGI*t*. The reason we refer to "digits" is because we usually refer to bits as being a "1" or "0", and because computers do their arithmetic in this strange kind of numbering system. This is because computers can only count from 0 to 1! We, of course, usually do our arithmetic in the base 10; that is, we use 10 symbols, 0 to 9, to do our ciphering. It is no doubt true that we use the base 10 because we have 10 fingers. Had we been born with 6, we probably would use a numbering system based on 6. Computers really have only two fingers—binary devices such as transistors, etc., and, therefore, calculate in the Base 2.

Characters Of course, we want to be able to represent decimal (Base 10) numbers and alphabetic symbols when we work with computers. In order to do this the binary numbers are *coded* to represent them. It happens that eight binary digits (called a "byte") can represent any decimal number and alphabetic character plus quite a few other symbols; in fact, 8 bits can represent a total of 128 *alphanumeric* symbols. A special code called "ASCII" has been agreed upon by international standards organizations for this purpose. A few examples of this follow:

Alphanumeric	Binary Code
A	01000001
5	00110101
a	01100001
@	01000000

Figure 10.2 illustrates the output of a device such as a personal computer sending a "5" to a remote location. The binary code 00110101 is output from the PC as a series of voltage levels representing the 0s and 1s.

Let us assume that -3 volts (DC) represents a "1" and 0 volts represents a "0". (This actually is not very arbitrary, as we shall see, since there has to be a "standard" agreed to by all device manufacturers hoping to interface with other data devices.)

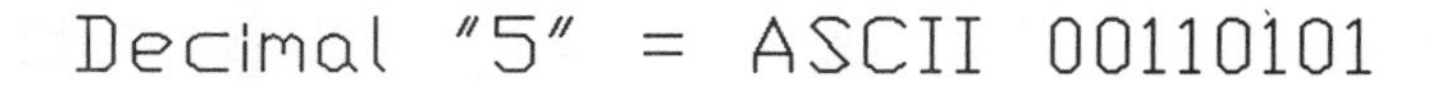

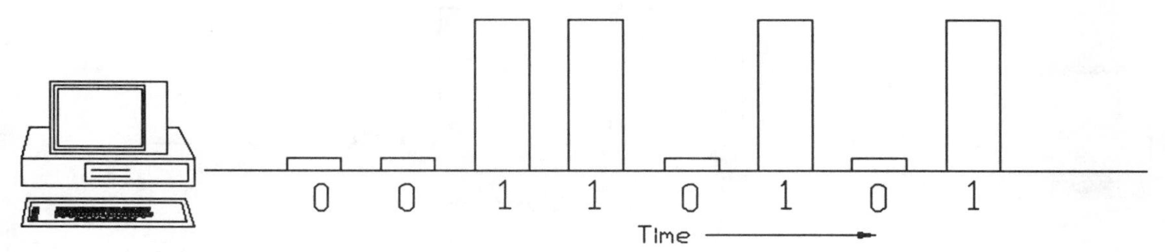

Figure 10.2 ASCII Code Example

The Problem It happens that relatively few of the circuits normally used for voice telephone transmission can successfully carry the binary (ON/OFF) pulses since they were engineered for optimum analog transmission. A major stumbling block is their inability to transmit direct current. So, in order to carry binary data signals, something must be done to convert the DC pulses into a suitable form for transmission.

The Solution Again, we can reach back in history to solve the problem. It happens that Alexander Graham Bell was working on a device called a "harmonic telegraph" when he conceived the telephone. The harmonic telegraph used vibrating reeds tuned to set frequencies to represent the dots and dashes. What if a particular tone or frequency was used to represent a "1" and another to represent a "0"? This is exactly what is done. See Figure 10.3, which shows an ASCII "5" converted to tones. A particular (analog) frequency is used to represent 0 and another for a 1 in the most elementary devices used to transmit data over phone lines. At the far end of the line, the process is reversed.

The frequencies must be carefully chosen to make certain that they will be carried by phone lines with the least *impairment* due to the line's characteristics.

Modem The device that converts the binary digits to tones and vice versa has come to be known as a *modem*. Modem is an acronym for MOdulate-DEModulate. There is a whole class of modems, ranging from rather simple devices that convert low speeds such as 110, 300, or 1200 bits per second to analog signals for transmission to complex devices that can transmit binary rates up to 19200 bits per second (19.2 kilobits per second).

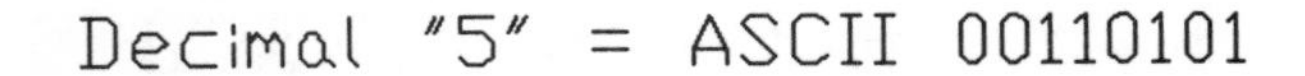

Figure 10.3 Conversion of Pulses to Tones

We will not examine the way that the very high speed modems work; suffice it to say that the technique used is to have each *change* of analog signal represent *groups* of bits rather than a single bit. The changes in the analog signal can be changes in amplitude, frequency or phase of the tones.

RS-232C Interface Figure 10.4 shows an end-to-end connection from terminal through modems and a phone line to another terminal. Notice in Figure 10.4 the notation of *RS-232C* at the interface to the terminals and the modems. RS-232C is Electrical Industries Association (EIA) "standard" developed to assure that there is an agreed-upon arrangement to interconnect data terminal equipment to data communications equipment. The standard is an American one, but there is also a similar international standard. The standard defines electrical and mechanical arrangements for interconnection. Recently, other standards have been agreed upon for the higher speed modems and for direct digital transmission to digital circuits.

Figure 10.4 Data Transmission Via Modems

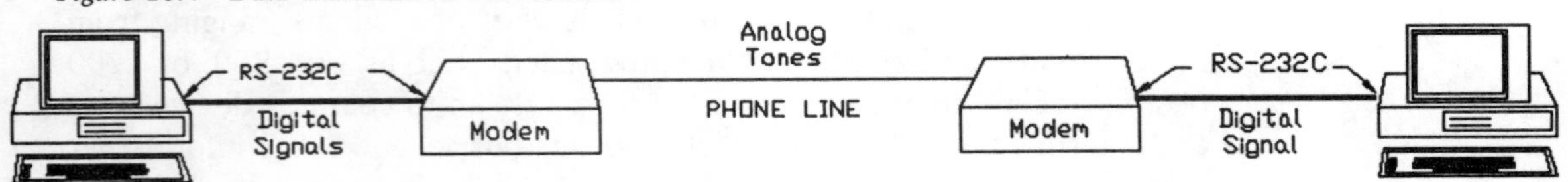

Bits and Bauds* The term *baud* is sometimes taken to be the same as "bits per second", but this is not always the case.

Referring again to Figure 10.3, it is apparent from the figure that each bit is converted to a tone that represents it. In this instance, a specific bit rate is the same as the baud rate. For the record, let us try to clarify the terminology:

> The term "baud" is a unit of signaling speed; the speed in bauds is equal to the number of signaling elements or symbols per second. Any of the three properties of a sine wave (amplitude, frequency or phase) can be made to carry signaling elements. For example, a tone put on a line can be changed in level, or volume, 1200 times per second, but each of the levels can be arranged to represent one bit of information. When more than two levels are used, each level represents a *group* of bits rather than just one bit. Four different and discreet levels can each represent two bits; thus, 1200 bauds can carry 2400 bits per second. In summary, baud rate represents the elements per second and each element can be changed in a manner to carry additional bits, giving higher bit per second rates than the baud rate. Stated another way, a baud can be thought of as a boat that carries bits as passengers—but there is a limit to the capacity of the boat.

The techniques used to squeeze more bits per baud are the subject of intensive research by modem manufacturers. Modems operating at 1200 baud are today carrying rates of 9600 bits per second, or higher. The fact is that there is not any modem that uses voice channels that operates at 9600 baud, but 9600 bits per second! This is likely to be in conflict with what many data vendors are apt to say.

Impairments If every telephone line was an ideal one, there would be no difficulty in achieving very high bit rates on every connection. However, there are some impairments to deal with on all but the shortest haul circuit. Some of them are:

Noise

Delay (so-called envelope delay)

Phase hits

Detailed discussion of these impairments is beyond the scope of this book. They can generally be accommodated by skillful design of modems. Special circuits can be ordered from the common carriers that are *condi-*

*The word "baud" is named in honor of a M. Baudot, author of one of the telegraph codes.

tioned to provide better characteristics. Failing this, another technique called "error correction" can be used to overcome transmission problems caused by impairments.

ERROR DETECTION AND CORRECTION

It is a good thing that human speech has a high degree of redundancy, because when you consider it, a telephone circuit does not faithfully reproduce speech. The bandpass of 300-3000 Hertz isn't close to the actual range of frequencies in speech. Furthermore, the English language usually demands that we use more words than are needed to convey meaning. All of us have a built-in capability to interpret and mentally correct for what is heard in speech, or the capability to detect and correct errors.

None of the data "languages" has built-in redundancy; however, it can be added before transmission to provide a means of detecting and correcting errors.

Odd or Even Parity Consider the ASCII code for the number "5" that we discussed earlier: 00110101. One technique that is commonly used to add redundancy to assist in the detection of errors is to decide ahead of time if the transmitted characters are to have an odd or even *parity*. As it stands, 00110101 has an even number of "1s". If an odd parity is planned, the extra bit would be a "1", resulting in 001101011 being transmitted. At the receiving end if the parity turned out to be even, then it would be known that an error occurred in transmission. But that's all. There would be no way of knowing how to correct that error. All that could be done would be to request a retransmission. This is a simple scheme, but it is used quite commonly.

Cyclic Redundancy Codes There are occasions when simple parity arrangements are not reliable enough (for example, if two errors should occur, the parity check would pass). Other schemes make use of several additional bits added before transmission that use algorithms called "cyclic redundancy codes" that can detect multiple errors, and can correct them as well. The price paid is the need to transmit many additional bits per character or group of bits.

Transmission Direction So far in our discussion of data we have not mentioned another important concept; that is, whether data transmission is to be one or two-way, or both simultaneously. Transmission which occurs only in one direction at a time is called *simplex*. Usually in this case, the entire bandwidth of a circuit is used to carry the modulated signal; hence, double the bit rate is available. For example, 9600 bps could be transmitted instead of 4800 bps, all else being equal.

Duplex transmission permits simultaneous transmission in both directions. Duplex is sometimes referred to as *full* duplex; however this appears to be redundant. Duplex transmission is virtually a requirement for applications such as a personal computer communicating with on-line database services or electronic mail services where there is a two-way interplay and very high bit transfer rates aren't needed.

TERMINALS

Probably the simplest data terminal was the telegraph key used in the early days of telegraphy. There has certainly been a remarkable evolution of terminals since then. We will examine a few of the more important ones:

Teletypewriters The teletypewriter was a device that was aptly named since it permitted direct typing of messages into a telegraph circuit. It was not only used for telegraphy, but was the initial terminal used for modern data transmission. But, of course, it had to be used with a modem when regular telephone lines were to be utilized. The use of teletypewriters (sometimes erroneously called "teletypes") has fallen off somewhat in North America, but heavy use is still made of them in the rest of the world for *telex* service.

Cathode Ray Tubes Cathode Ray Tube (CRT) terminals came into popular use as more need occurred to communicate with a computer main frame directly. The really big emphasis came from the airline industry. The primary advantages offered by CRTs are high speed and when paper output is not required. In virtually every application of CRTs, communications are with a main frame "front end", or communications controller.

Personal Computers Almost everyone is familiar with the astounding growth in the use of personal computers, and many of them also have modems that are used for bulletin boards, electronic mail, and downloading of software. It is highly unlikely that anyone saw the proliferation of modems that has occurred, and they have been greatly reduced in physical size. By far most of the modems for this use are limited to 1200 bps operation, but there has been a migration toward 2400 bps.

OTHER DATA SERVICES

Telex and TWX Telex and TWX services predate other data transmission services, since they were an outgrowth of telegraphy. It is a written message service. In the 1930-40s many U.S. businesses subscribed to TWX, "teletypewriter exchange service", while their counterparts in the rest of the world used virtually the same service, but called it "telex". The latter are still very heavy users of the service, which is generally provided by the post office.

11 DATA TRANSMISSION II

In Chapter 10 we established some fundamental concepts of data communications. This chapter will carry some of the concepts a little further.

There are many instances where a group of terminals that are colocated have need to send and receive data to a remote point. We will look at some of the ways to accomplish this without devoting a telephone line to each of the terminals. The way this is done is to use a technique we looked at earlier—multiplexing.

Simple Multiplexer Refer to Figure 11.1. We see four terminals at the left, each operating at 300 bps. Instead of each of them being connected to a modem, each is terminated in a device called a *multiplexer*, or "mux", as it is sometimes called. The four 300 bps signals are combined as a 1200 bps composite signal by the mux. After the signals are combined the 1200 bps can be put through a modem for transmission on a telephone line. Note that this concept can be expanded to handle many more low-speed signals in various combinations, just so the sum of the

Figure 11.1 Simple Multiplexer

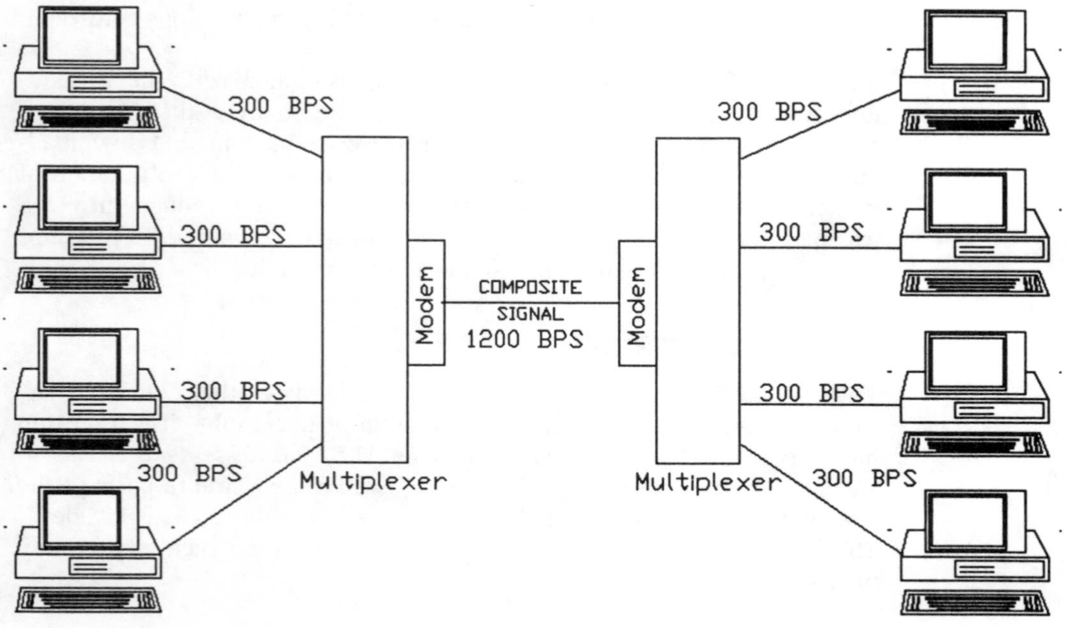

inputs does not exceed the mux bit rate; for example, four 1200 bps and two 2400 bps signals could be input to a mux operating at 9600 bps.

Statistical Multiplexer In some applications not all terminals are actively sending or receiving data at *exactly* the same time. It is then possible to have a total number of possible inputs exceed the mux speed. A special type of mux that uses time division and buffering is required for this kind of operation. They are called *statistical muxes* because they take advantage of the statistical characteristics of the terminal activity. Refer to Figure 11.2 for an insight into their operation.

Notice that each of the terminals is operating at a rate of 1800 bits per second, but that the signal presented to the line is only 4800 bits per second. This appears to be an anomaly since the product of four 1800 bps signals is 5400 bps; however, the "stat mux" has three things going for it that permit such operation: 1) the output of the mux is a steady 4800 bps and the inputs are assumed to be sporadic; 2) if the inputs should exceed the output rate of the mux, a buffer will temporarily absorb the overflow; 3) the mux can command the terminals to halt transmission for a period. Obviously, the system traffic characteristics must be carefully examined before settling on the *concentration* for the system.

In this figure we see that the output of the distant stat mux is being fed

Figure 11.2 Statistical Multiplexer

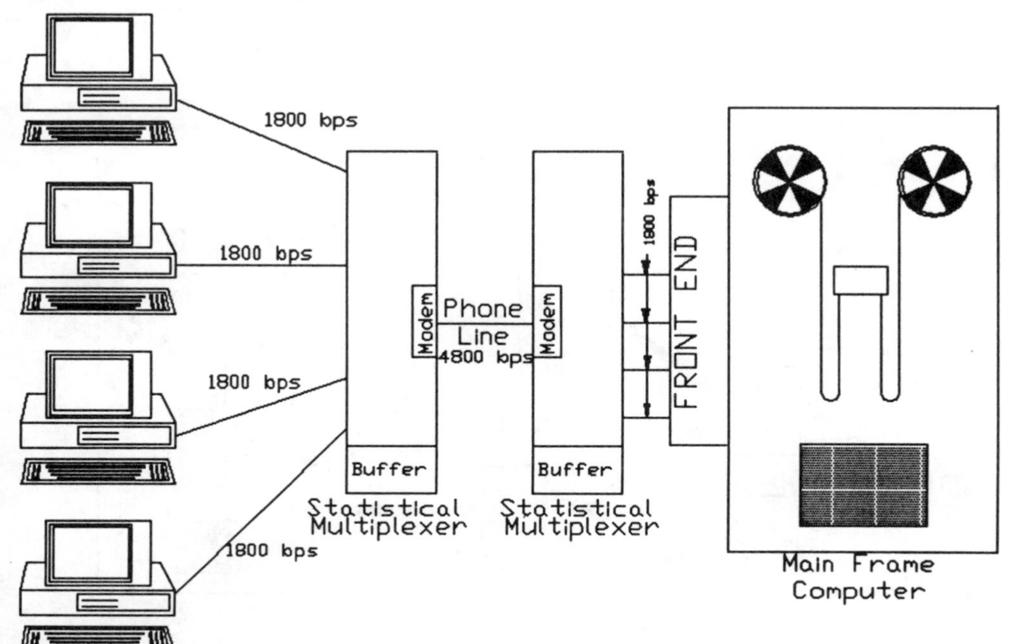

to the "front end" of a computer. Although the term "front end" is commonly used, the proper name for this unit is *communications controller.* The communications controller is a main frame computer's window to the world of communications. Some communications controllers have the capability to replace the stat mux equipment at its end of the system; in this case the incoming demodulated line signal from the modem goes directly into the front end of the main frame computer.

Packet Switched Data Another technique that has gained a great deal of favor with providers of services to personal computer users, (such as CompuServe, Source, Dow Jones, etc.) is called "packet switching". The reason that packet switching is so popular is that the intercity circuits are not tied up for the duration of the caller's use, with a great deal of time wasted while the user is thinking or typing at a relatively low speed. Instead, a "virtual" connection is established which only uses the intercity facilities when there is actual data to be transferred. Figure 11.3 illustrates this technique.

As was previously the case, PC users dial a local telephone number to gain access to the system; however, at this point the arrangement is quite different. The receiving modems are connected to a "node", which is a gathering point for local users. Nodes are either connected directly to the system's central processor host or are connected to another distant node. Nodes are interconnected via high-speed digital circuits.

Figure 11.3 Packet Switching

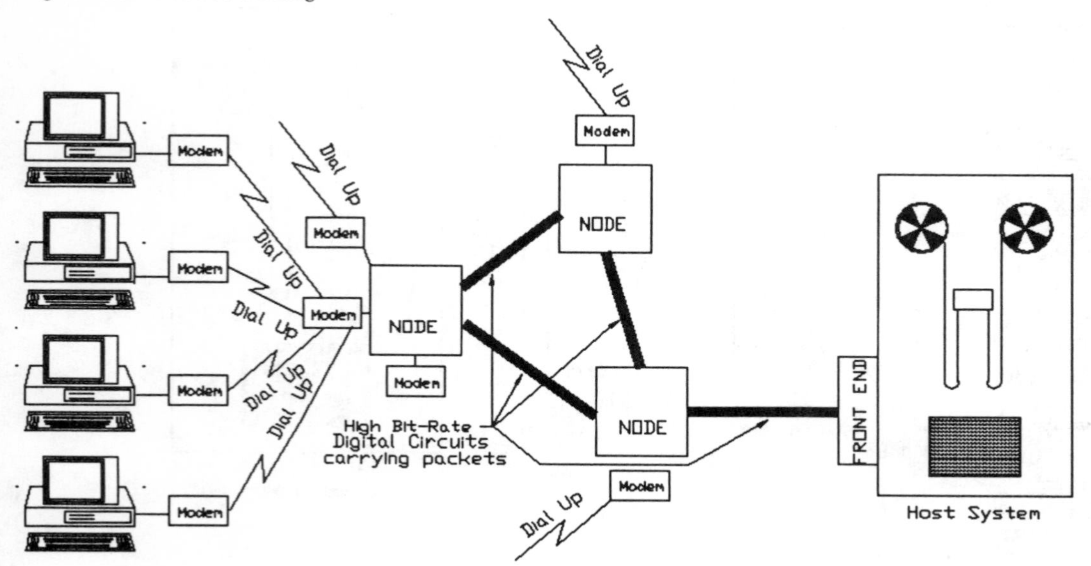

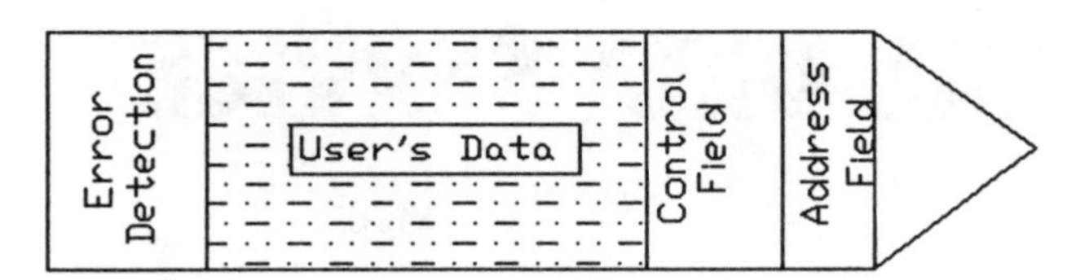

Figure 11.4
Format of a Packet

Several things take place in a node:

A "virtual" connection is established from the end user to the host system as needed.

Data from the user or host is stored briefly.

The data is "wrapped" in error correction protocol and attaches an "address" that it has stored for the connection. See Figure 11.4.

The combined data is transmitted as a "packet" to another node or to the host.

When data from a distant node receives a packet it checks for transmission errors. If an error is detected it requests a retransmission from the originating node.

At the terminating end the protocol information is stripped from the packet and the user's data is passed to the host (or PC).

12 DIGITAL DATA SERVICE

The advent of digital carrier systems, such as T1 discussed in Chapter 4, has been of great interest to the larger users of data transmission. Since both the source of data signals and the medium used to carry them are digital, there is a natural fit. In recent years there has been a proliferation of digital transmission, so there has been increased interest by the users in gaining direct access to these circuits; that is, without the use of modems. Today, interfaces exist to make this possible and their applications are increasing.

Recalling from Chapter 4 that each voice channel in T1 carrier systems is actually a 64 kilobits per second digital signal, it would seem that very high-speed data streams could be carried. This is the case. Where the use of modems constrained data transmission to 9.6 kilobits per second or so a basic T1 voice channel can handle 64 kb/s. What is needed to accomplish this is:

A means to "deliver" the digital bit stream to the user's premises.

A standard interface to the user's terminal equipment.

A means to control (or dial up) the connection.

Figure 12.1 illustrates the use of one of the 64 KB/S time slots for data transmission. The other 23 are used for voice channels.

ISDN Recognizing the need for a standard way to provide high-speed digital service, an international standards committee was commissioned to define the way. What resulted was a system called "Integrated Services Digital Network", or ISDN.

ISDN is a system that provides simultaneous voice and high-speed data transmission through a single channel to the user's premises. A special computer-controlled switch in the central office transforms ordinary copper wire pairs to having the capability of carrying the signals.

Figure 12.2 illustrates the two basic systems that were standardized:

2B + D Provides for two 64 kb/s channels to carry voice or data plus a 16 kb/s signaling channel.

23B + D Provides for 23 64 kb/s channels and a 64 kb/s signaling channel.

The "B" channels are used to carry the data information that is to be transmitted, while the "D" channel carries addressing and other call-related information.

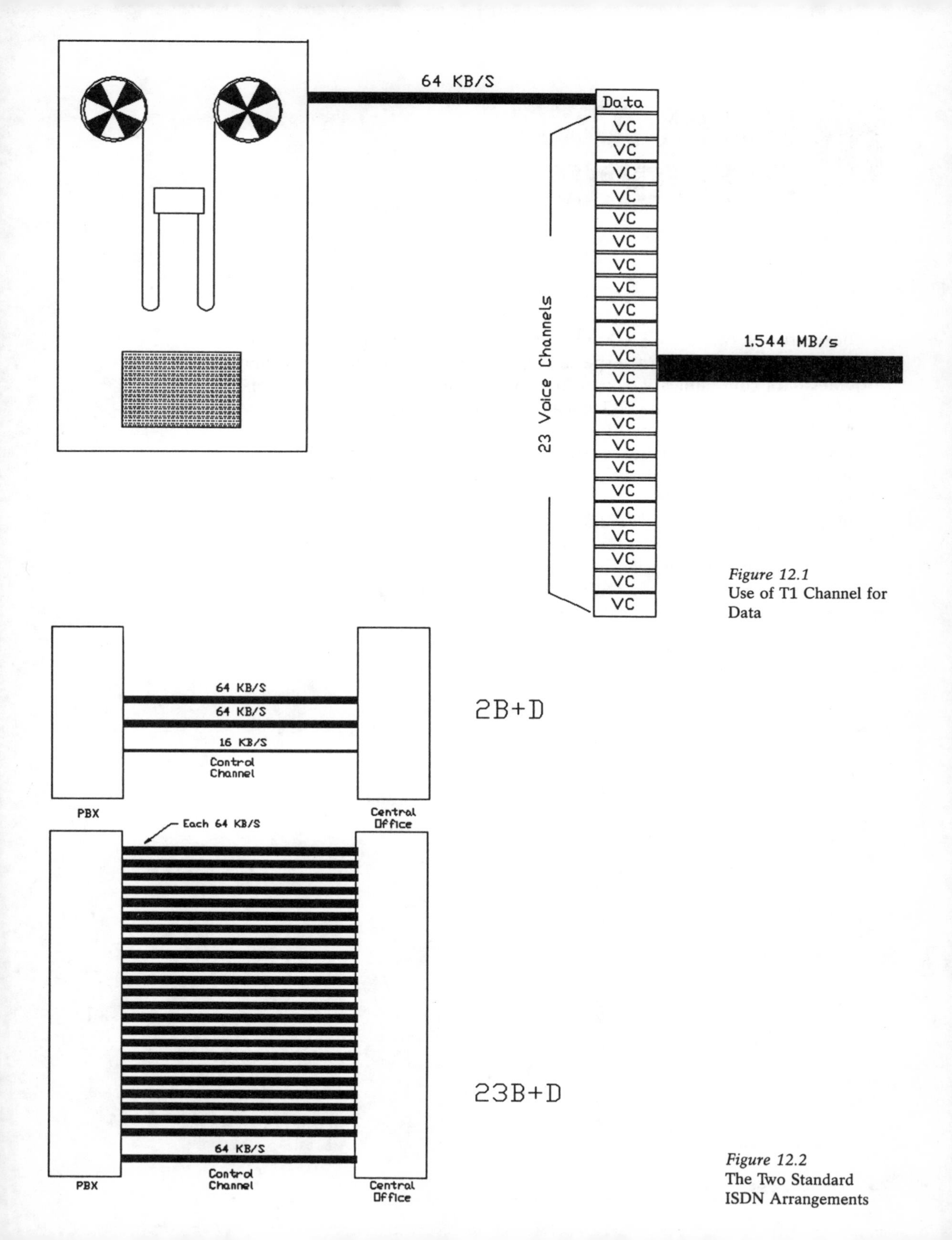

Figure 12.1
Use of T1 Channel for Data

Figure 12.2
The Two Standard ISDN Arrangements

13 LOCAL AREA NETWORKS

Local Area Networks (LANs) are a recent development in the telecommunications arena. LANs are a special, high-speed dedicated network that provides data communications capability within an office or group of offices in a campus environment. Typically, a "drop" is provided at all likely worker locations. See Figure 13.1.

Personal computers have been the driving force behind the development of Local Area Networks—LANs; a typical modern office has an abundance of PCs. However, it is not only PCs that have spurred the development, but the need to link copiers, facsimile, plotters, printers, etc. together. Also, devices sometimes called "file servers" that provide mass storage for PC users can be connected to the LAN. An illustration of such an arrangement is shown in Figure 13.2. The diagram also shows that an extension to the "outside world" by means of a *gateway* or modem pools can be arranged.

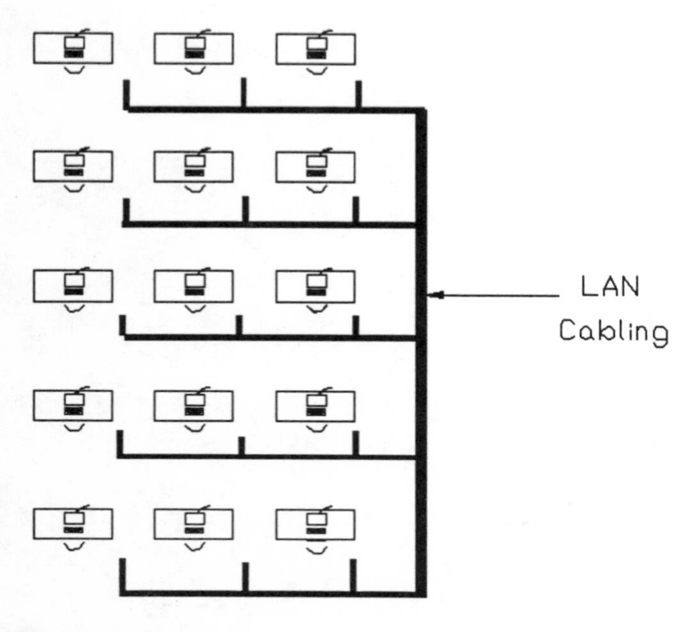

Figure 13.1
An Office Arranged for
A Local Area Network

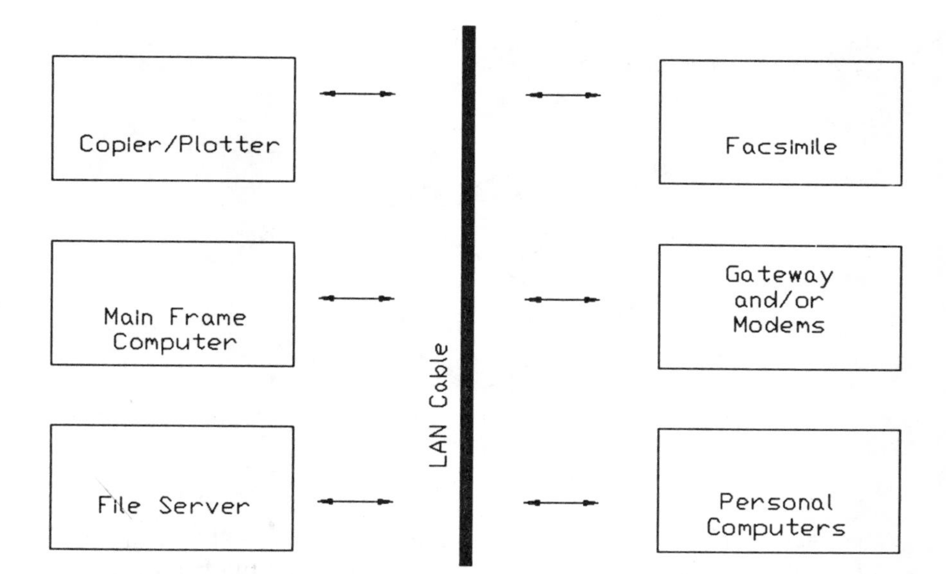

Figure 13.2
Examples of LAN Usage

Configurations Several configurations of LANs are in use. Figures 13.3 through 13.6 illustrate them. The actual cable facilities that can be utilized wil be discussed later.

Star cabling (Figure 13.3) is very much like the cabling provided for telephone hookups. Each station is provided with a "home run" to the LAN Center where some form of switching is provided to interconnect the desired devices.

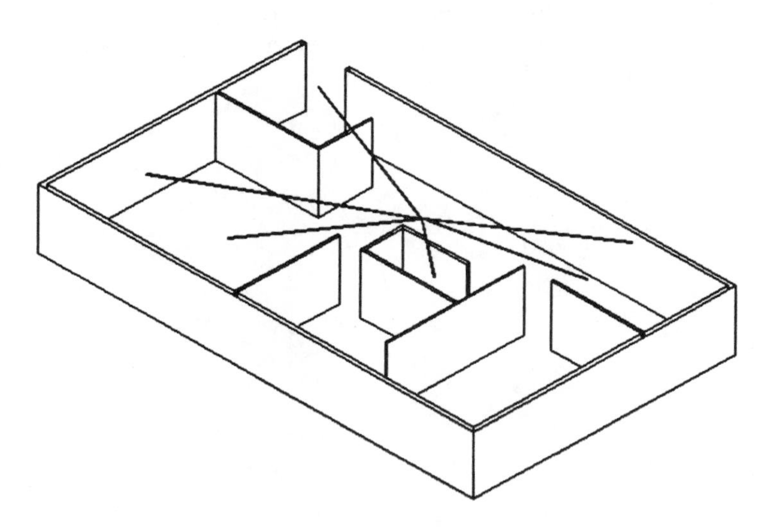

Figure 13.3
Star LAN Layout

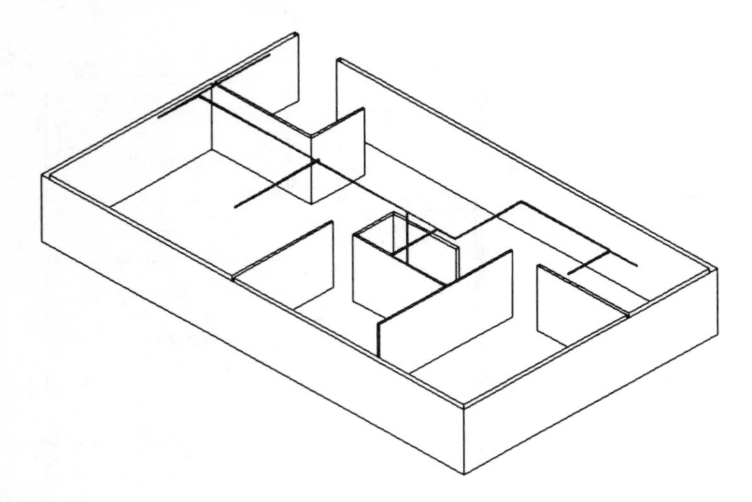

Figure 13.4
Tree LAN Layout

Tree cabling (Figure 13.4) uses some form of wideband cable with drop-off points as needed. Each drop is provided with the full bandwidth or data rate of the cable.

Bus cabling (Figure 13.5) is routed so that each station "sees" all of the signals that are broadcast on the cable.

Ring cabling (Figure 13.6). This arrangement is sometimes called "bucket" or token passing. Data originated at a station is inserted before passing the package on.

Access Methods Depending on the configuration and other factors, LANs can either use contention or deterministic methods for network access. The most common form is contention where the station determines when it is to transmit data. Methods of avoiding collisions have been developed, since they are quite possible. Deterministic is a polling or assignment means of calling for transmission.

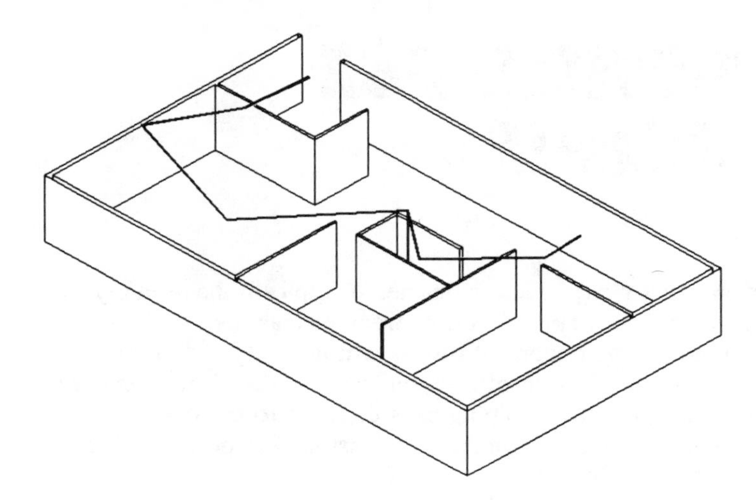

Figure 13.5
Bus LAN Layout

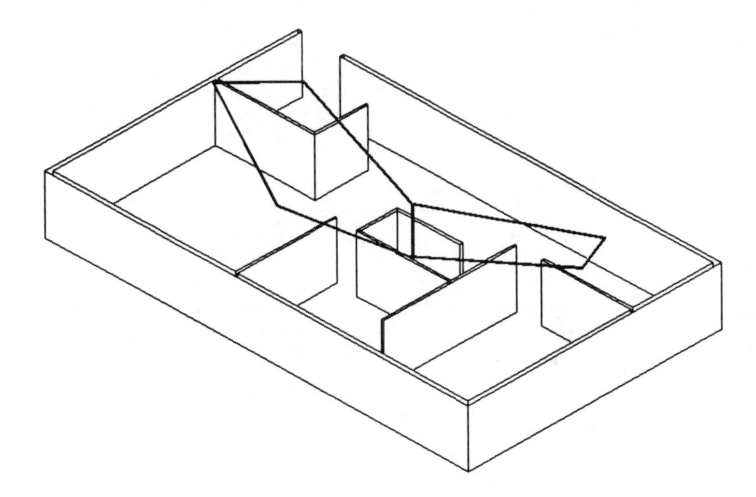

Figure 13.6
Ring LAN Layout

14 ADVANCED PBX CONCEPTS

Centralized Answering Larger businesses typically have many offices that are geographically spread out. In areas such as the Northeastern United States, there can be dozens of offices within a fifty-mile radius, or so. Also, typically, each of these offices will have a PBX, and someone must be assigned to attend to incoming calls directed to the operator position. Frequently, this task is assigned to the least experienced employee. This can be a poor idea. Customers calling in can be mishandled due to a lack of knowledge of people, products and services on the part of the attendant. Centralized answering system (CAS) is an effective solution to this problem. Refer to Figure 14.1.

Figure 14.1 Illustration of Centralized Answering

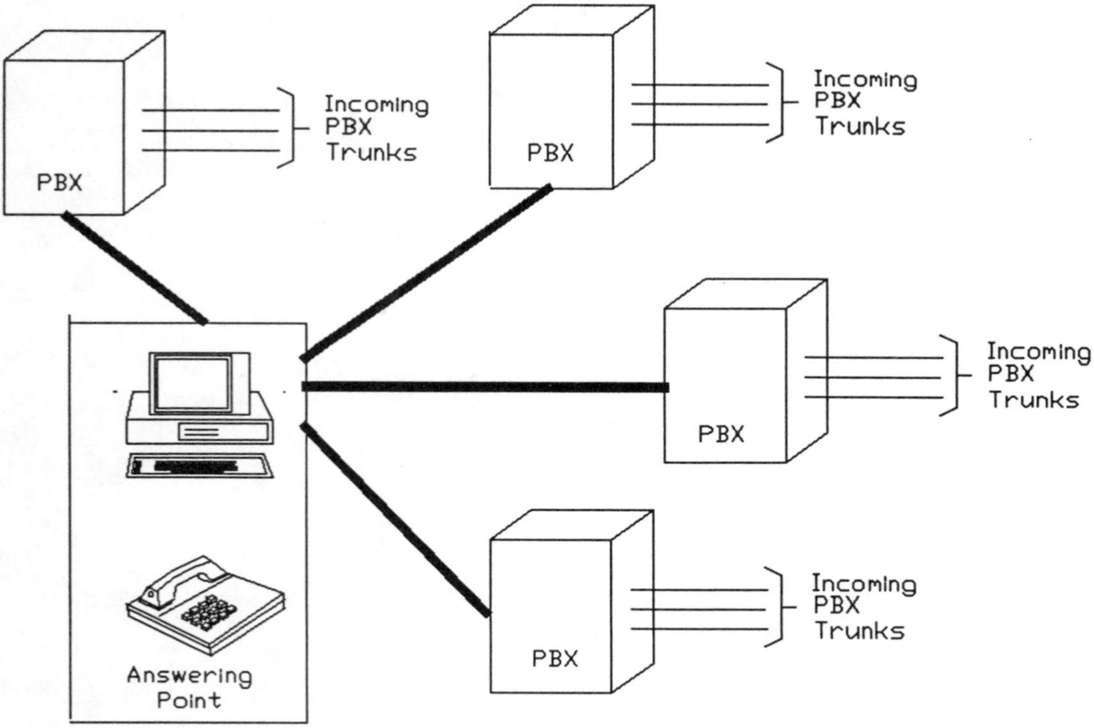

The essential elements of CAS are:

Incoming trunk calls at each PBX are temporarily connected to a remote central location.

A data link is provided between each PBX and the central location.

An easily accessed database (usually computerized) is provided at the central location for the attendant to have intelligent information to handle calls.

Operation of CAS is as follows:

Incoming calls to a PBX are routed to the CAS center, where a trained attendant answers them. The attendant is made aware of the identity of the PBX automatically. Frequently, the incoming calls are connected to the attendant via an automatic call distributor. (See ACD below).

Attendant determines the disposition of the call, referring to database as necessary to obtain correct PBX extension or other call routing information, and . . .

Remotely causes the distant PBX to connect the call to the proper extension. If the call cannot be completed due to busy or no answer, the attendant is made aware of the difficulty so that the caller can be informed. The call can be "parked" and released from the attendant, to be returned after an interval for the attendant to retry the connection.

Call Distributors Call distributors are devices that do exactly what the name implies; they distribute incoming calls to an establishment in such a way that the callers are treated equally. (See Figure 14.2). We have all experienced call distributors on calls made to airlines, service organizations, etc. There are two general classes of call distributors:

Uniform call distributor (UCD)

Automatic call distributor (ACD)

Both types provide the function of evenly distributing incoming calls to available *agents* (as they are usually called in this application). Other functions of both types are:

Hold calls in a queue for completion

Provide "music on old"

Provide call counts

UCDs are usually simpler devices than ACDs and are frequently found as a feature of a modern PBX.

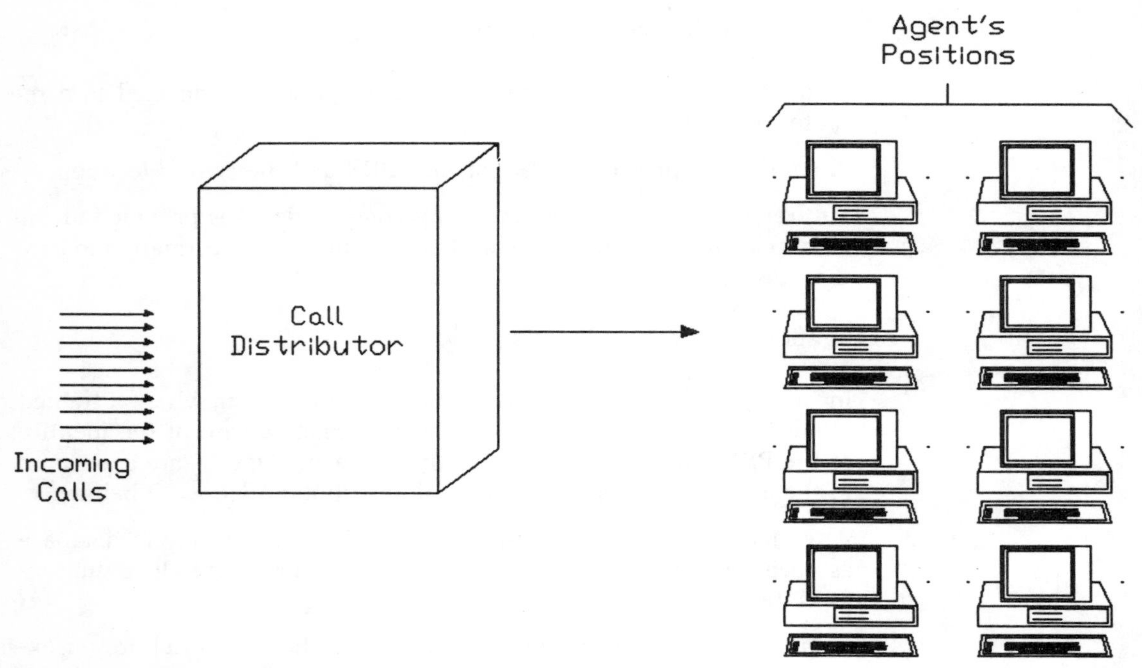

Figure 14.2 Automatic Call Distribution

In addition to the features of an UCD, the processor-equipped ACDs, which frequently are stand-alone devices, provide statistical information and processing capability such as:

- Number of calls waiting at given times,
- Information on how each agent handles calls; e.g., time to handle callers, processing or "work" time, etc.,
- Retain daily statistics for further analysis.

ACDs also provide a means for an agent to halt incoming calls while processing information from previous call, and is known as "work time".

Least Cost Routing Larger PBX users frequently establish networks that make it desirable to provide an automatic means to select the best way to route calls. Referring to Figure 14.3, we see a PBX that has a number of paths to the "outside world". The selection of trunks illustrated is only representative, but it isn't unusual. Frequently, a large PBX will have even more choices of routes.

In earlier days, it would be up to the users or the attendant to select what seemed to be the most economical route and dial the (many) digits

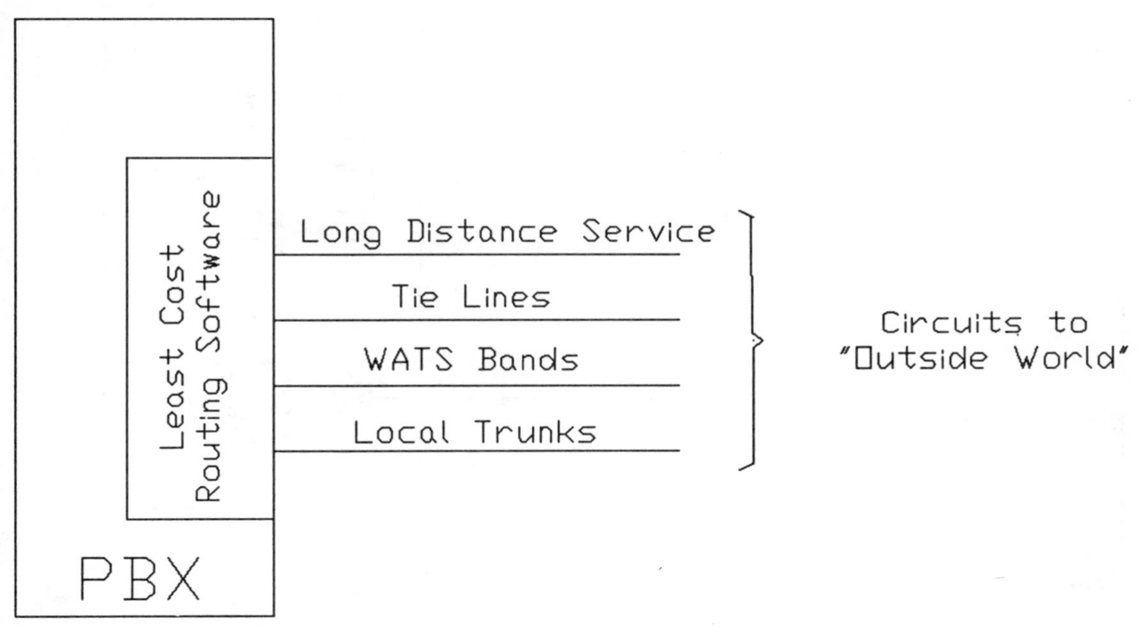

Figure 14.3 Least Cost Routing

necessary to access the facilities. Left up to the typical user, he would probably select the easiest (and probably most costly) way to complete a call.

Least Cost Routing (LCR), also called *Automatic Route Selection,* software in a PBX provides the capability to take over the tasks of selecting the best routing for a call and inserting or deleting the necessary digits to satisfy the requirements of the service. If all of the primary circuits are busy, the call can be automatically routed to a more costly route, optionally based upon the class of service assigned to the caller.

Simultaneous Voice and Data The "digitization" of PBXs has brought an important new capability to PBXs; the ability to provide stations on the PBX that can simultaneously carry data signals and voice. In this way, a user can be using his PBX extension to make and receive calls at the same time that he is using, for example, a personal computer connected to a central database. The same wires that provide voice service are carrying *dial-up* data signals. See Figure 14.4 (next page).

This capability is feasible because the switching matrix is using binary data-like signals to carry voice signals. Digital data signals are essentially the same as digitized voice, so the matrix carries them in the same manner. The voice and data signals are split at the PBX and carried independently through the switch. The station line circuits and station sets are special and are generally proprietary.

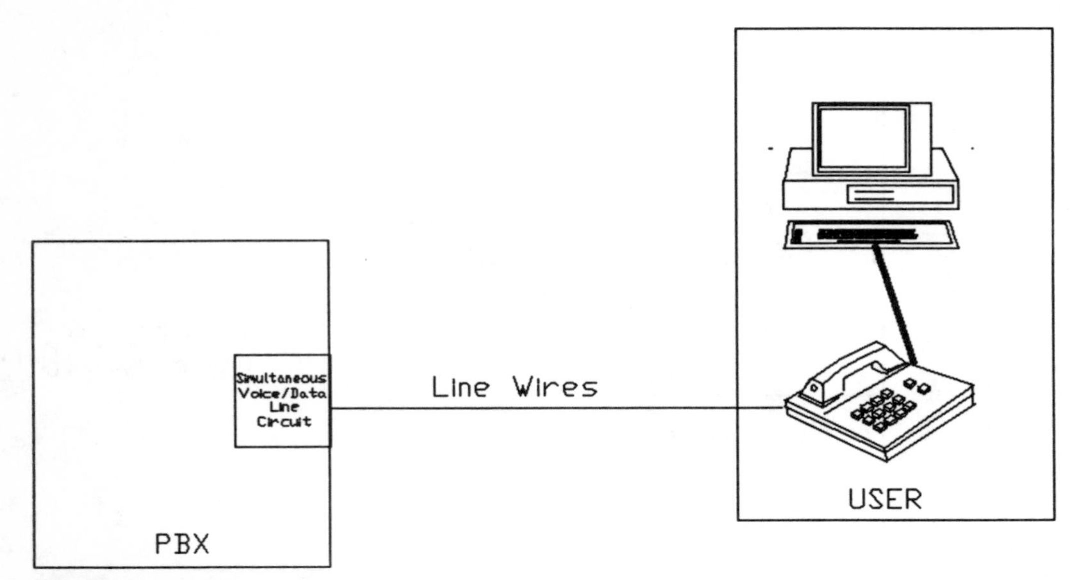

Figure 14.4 Simultaneous Voice/Data Station

Departmental Calling In Chapter 5 we discussed Direct Inward *Dialing,* a service that enables callers to dial directly to a given station behind the PBX. Where DID is not provided it is still possible for callers to be able to select a specific group of stations, for example, the service department of a firm using Departmental Calling. A dedicated group of incoming trunks is automatically connected by the PBX to the employees (agents) dedicated to the particular function. See Figure 14.5.

Usually, the stations are part of a "hunting group" or in "rotary". Calls to these stations can be transferred to any other station in the PBX. Also, it is not unusual for the incoming calls to be controlled by a call distributor.

Voice Messaging Everyone who has worked in the office of a large firm is familiar with the "game" of telephone tag. Millions of the little pink or yellow slips are probably used every year. DTMF (tone) phones in conjunction with computer-controlled devices arranged to provide "voice messaging" service are making it possible to eliminate this annoyance. Referring to Figure 14.6, we see how they work.

Basically, a VM unit stores a digitized message that the caller leaves when his call is not answered. Upon his return, the person called calls the VM unit to receive his messages. Using his DTMF dial pad he is able to control the VM unit to provide a number of functions:

Determine the number of messages stored for him.

Individually select the message(s) desired.

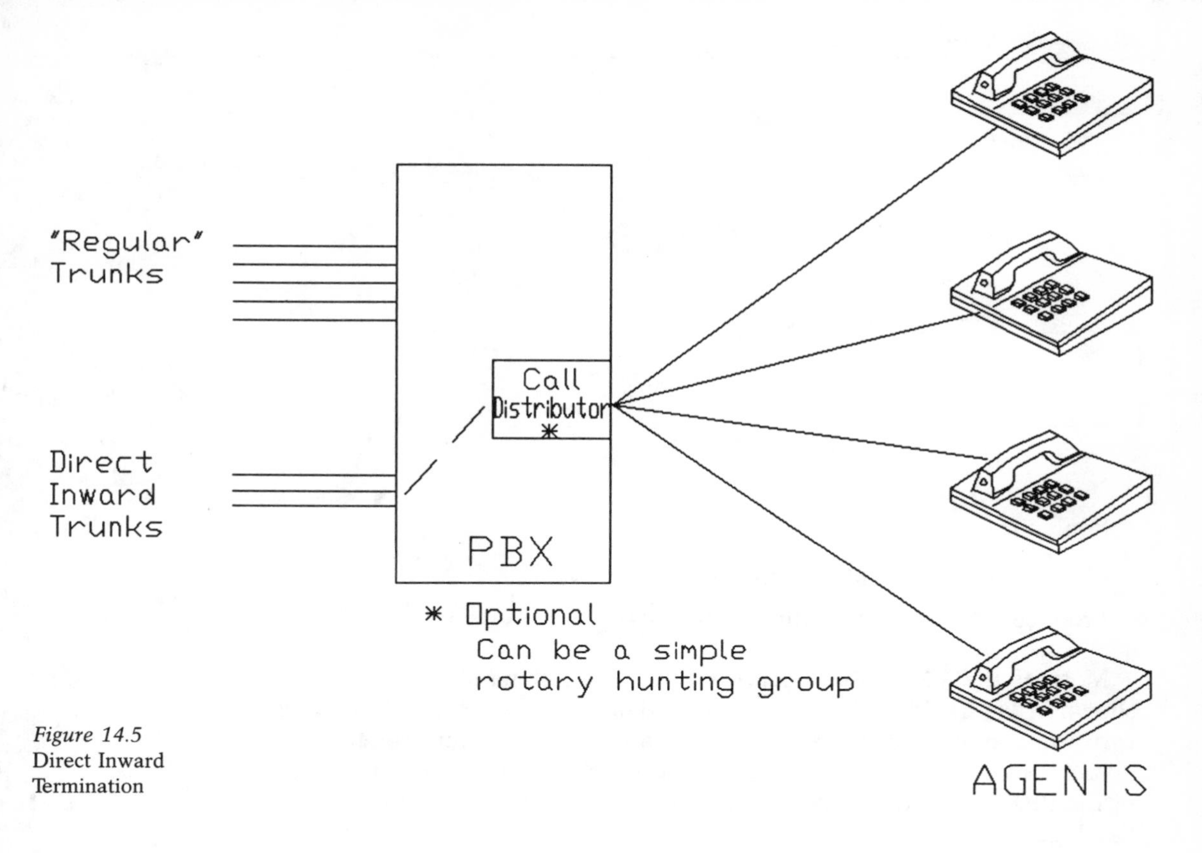

Figure 14.5
Direct Inward
Termination

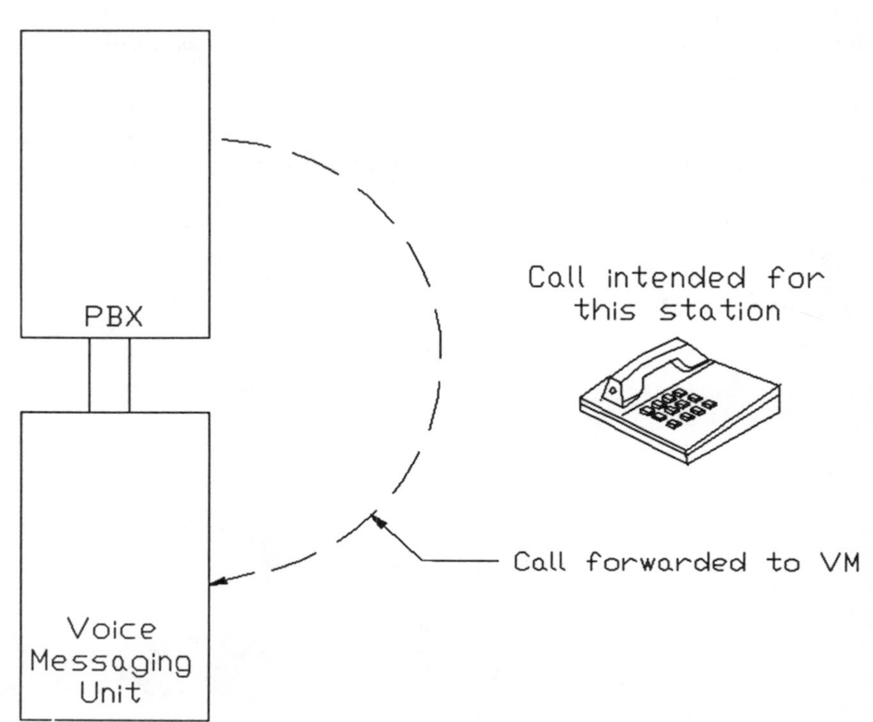

Figure 14.6
External Voice Message
Equipment

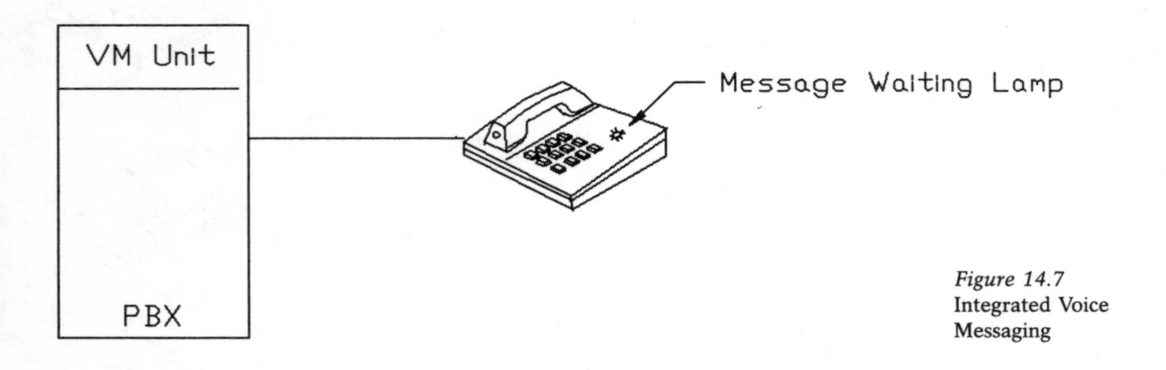

Figure 14.7
Integrated Voice Messaging

Replay messages.

Store his own message in response to a message.

Produce messages for multiple recipients.

Most of the VM units in service to date are stand-alone devices, as shown in Figure 14.6. A good way to arrange the system is to have calls forwarded automatically to the VM unit after a "no answer" condition.

More and more PBX manufacturers are including the function as part of the PBX, or alternatively, arranging for direct interaction with the VM. See Figure 14.7.

An important function that results is an indication to the users that a message is indeed awaiting his attention. This is known as a call waiting indicator.

Information about the author of this book follows on page 91.

INDEX

ABOUT THE AUTHOR . . .

Jack Dempsey is just the person to write a book on basic telecommunications. His interests, professional experience and wide array of technical skills and knowledge are the background for this book. He's also a master teacher who writes logically and understandably and knows how to uncomplicate the always technically complicated world of telecommunications.

Mr. Dempsey has an impressive career that spans 35 years in the telecom field. After receiving his Bachelor of Science Degree, Communications Major, from the University of Illinois, he joined IBM as a Customer Engineer. Following this, he worked at Illinois Bell in a number of rotational assignments mostly related to #5 Crossbar. After an assignment in IBT Transmission Engineering, Mr. Dempsey was transferred to Bell Labs as a member of the technical staff where he further developed #5 Crossbar common control. While on this assignment, he conceived an invention which was later awarded a patent.

Mr. Dempsey also worked at Western Electric and later became a faculty member of the esteemed Cooperstown Data Communications Training Program. He held teaching positions in marketing, engineering and plant curricula. He was then transferred to AT&T Headquarters, Data Engineering Section. While there, he was primarily involved in wideband transmission design. After this period with the Bell System, he left to join Xerox.

His initial assignments at Xerox were with the Communications Product Div., Engineering Group, and he was appointed Chairman of EIA TR 30.2, the committee that produced the Data Interface Standard, RS-232C. Work on this committee also included joint standards development efforts with ISO (International Standards Organization). Mr. Dempsey participated as a member of the U.S. delegation to CCITT in Geneva a number of times. He also served as Chairman of an FCC Advisory Committee with the task of devising an acceptable arrangement for direct connection of non-Bell hardware to Bell lines.

He was one of the founding fathers of EPSCS, Bell's enhanced CCSA-like network. During its design phase he worked cooperatively with AT&T in defining this service.

Dempsey was a member of the Telecommunications Dept. of Xerox for seven of his fourteen years with that company. His responsibilities encompassed both voice and data communications. His last assignment with Xerox was as manager of a program to replace Xerox' extensive, outdated message system. A unique system built around a message switching computer and the Xerox EPSCS voice network was implemented. This project resulted in a savings to Xerox of over $2 million per year.

Most recently, Mr. Dempsey was Senior National Consultant to Interline, a subsidiary of U S West. In this capacity he was called upon to provide Interline with consultation on a diversity of topics related to telecommunications. In addition, he conducted a number of training sessions.

When Jack Dempsey isn't writing books, he relaxes with this wife Barbara. He divides his retirement time between Florida and the Canadian wilderness. His hobbies are telephony, video photography and his personal computer which is always by his side.